MECHANICAL MAN

DEAN E. WOOLDRIDGE

Research Associate
California Institute of Technology

MECHANICAL MAN
The Physical Basis of Intelligent Life

McGRAW-HILL BOOK COMPANY

New York   San Francisco   Toronto   London   Sydney

MECHANICAL MAN

1234567890VBVB7543210698

Acknowledgments

I have been fortunate in being able to elicit from a number of very busy scientists comments and suggestions that have helped me in the preparation of this book. Without implying that these men necessarily endorse what I have written, but only for the purpose of gratefully acknowledging their assistance, I list their names here: J. B. Angell, Stanford University; H. L. Dreyfus, Massachusetts Institute of Technology; W. V. Houston, Rice University; J. R. Pierce, Bell Telephone Laboratories; J. B. Platt, Harvey Mudd College; J. A. Robinson, Rice University; A. L. Samuel, Stanford University; J. C. Shaw, Rand Corporation; J. R. Whinnery, University of California at Berkeley; and R. J. White, Western Reserve University.

Dean E. Wooldridge

Contents

PART 3: INTELLIGENCE

PART 4: CONSCIOUSNESS

Introduction: The Nature of
Physical Explanation

This is essentially a report on modern attempts to account for the origin and properties of living organisms, including man, by means of the principles of physics. Such attempts are largely inspired by the impressive past successes of physical science in transferring one after another of the details of human experience out of the supernatural into the realm of the understandable. Rain and wind, lightning and earthquake, and the rising and setting of the sun and stars have long since been accepted as manifestations of the workings of the laws of gravity, mechanics, thermodynamics, and electricity. In more modern times the propagation of radio waves, the properties of chemical dyes and plastics, the Van Allen belt, nuclear energy, and the principles of rocket propulsion are all "understood" in terms of generally accepted natural laws.

But there is another reason for the strong attraction that physics holds for scientists: its fundamental simplicity. To be sure, there is less optimism than there once was about the possibility of devising two or three general mathematical expressions containing a complete statement of all the physical principles that govern the universe or of discovering three or four basic sub-nuclear particles from which all matter is derived. Nevertheless, relative to its accomplishments, physics still works with surprisingly few basic ingredients. With a couple of dozen particles and a similar number of fundamental laws we are today able to derive explanations for a tremendous variety of physical and chemical phenomena, and most of those we cannot explain appear to be beyond our reach because of limitations of our analytic ability, rather than because of any inadequacy in the fundamental laws. It is true that the last word has not yet been said relative to the basic particles from which matter and energy are derived and that we have good reason to believe we have not yet precisely formulated the natural laws, since new discoveries require us to refine and restate them from time to time. We cannot even be sure that there do not still exist undiscovered physical phenomena whose explanation will require major additions to our present formulation of the body of natural law. However, all this is beside the point. The fact that our knowledge is less than complete cannot obscure the convincing evidence that all phenomena we call physical or chemical are obedient to the operation of a relatively small number of physical laws in matter containing a relatively small number of irreducible particle types.

And the appealing simplicity of physics extends to another of its features: its absolute orderliness. The laws of physics are never broken. In the realm of physical and chemical phenomena every detailed event, whether it is the formation of a new galaxy or the fall of a raindrop, is the lawful effect of causes which are themselves the lawful effects of other causes, and so on, going back ultimately to the fundamental particles and the basic laws of the universe. Even the existence among the laws of a principle of indeterminacy limiting the precision with which the future can be predicted does not permit the entry of caprice into the world of the physical scientist. Within a calculable and frequently very narrow range of uncertainty, the future is completely determined by the past. Given the laws and the particles, all else follows inexorably.

Undoubtedly it was an incipient awareness of these appealing aspects of physical science that led, in the seventeenth and eighteenth centuries, to attempts to apply it to biology. As we shall see, some of these applications were indeed successful in explaining mysteries of life that had long been thought to lie forever outside the range of phenomena susceptible to human understanding. Inevitably, these early successes stimulated further search for unsolved problems which might yield to the methods and principles of physics. And it was gradually discovered that this approach to biology became steadily more fruitful as physical research led to improved understanding of the natural laws underlying mechanical, electrical, atomic, and molecular phenomena.

Indeed, with the passage of time, the discovery of solutions for long-standing and difficult problems of biology by application of the principles of physics developed into one of the most persistent trends in the history of science. So rapid did this trend finally become, and so spectacular are some of the related modern discoveries, that many scientists have now concluded that there is no fundamental difference between biology and physics—that ultimately *all* aspects of the structure and behavior of living matter will be explainable in terms of exactly the same particles and natural laws as those that we find underlying the load-carrying qualities of a bridge, the flight capabilities of a rocketship, or the color of a sunset. According to this magnificently unifying concept, there is but one ultimate science, and that is the science of the physicist.

Hence the preoccupation of this book with the thesis that all biological phenomena have physical explanations accords with a dominant theme of modern scientific research. Nevertheless, an attempt is made not to prejudge the issue. Indeed, until late in the book when the evidence has all been examined, the thesis is used primarily as a literary organizing principle for the structuring of a discussion of many of the exciting recent discoveries related to the science of life. Thus, on the way to a physical explanation of biology, we shall concern ourselves not only with modern knowledge related to the origin and so-called physical properties of organisms but also with that related to such "nonphysical" properties as behavior, intelligence, and consciousness. The resulting case for the reducibility of biology to physics may not be convincing to all. Nevertheless, the subject matter treated, comprising as it does a modern view of the origin and principles of operation of organisms including human beings, should be interesting to most.

I have dealt with similar subject matter in two previous books, *The Machinery of the Brain* (McGraw-Hill, 1963) and *The Machinery of Life* (McGraw-Hill, 1966). Indeed, approximately half of the material of this book—that of Chapters 1 through 6 and 13 through 15—has already been covered in these earlier works. However, most of this material has been shortened and simplified to make it more easily readable by nonscientists. And Chapters 7 through 11, on intelligence, are all new. So are Chapters 16 through 19, on the implications of the physical explanation of biology. Incidentally, these last chapters carry the discussion into matters, such as the nature of personality and free will, that have been traditionally considered more appropriate for the philosopher than the scientist to deal with. I invade this realm only because if the inferences that many scientists are drawing from the results of modern biological research are correct, then some of these so-called philosophic concepts also qualify for at least limited treatment by the ordinary methods of physical science. As befits the essentially nonphilosophic orientation of the book, the terminology and method of approach remain, in these closing chapters, those of the physicist or layman rather than those of the philosopher.

Our first objective will be an explanation of the "physical" properties of plants and animals and of how living organisms possessing such properties came into existence.

part 1

The Physical Properties
of Organisms

chapter 1

The Chemistry of Life

The first evidence that some of the properties of living organisms might be susceptible to physical explanation appeared early in the history of science. Indeed, the event that caused the title "father of modern biology" to be bestowed upon William Harvey was his discovery, in the early 1600s, that the ordinary engineering principles that govern the pumping and flow of liquids are capable of accounting for the functions performed by the heart, an organ that had previously been thought to belong in the realm of the unknowable.

Since Harvey's time, application to the body organs of the principles of physical science has led to many useful discoveries. In addition to knowing that the heart is a pump, we now know that the lungs comprise a mechanism for the introduction of oxygen into the body's chemical plant and for the extraction of gaseous waste products; we understand a great deal about the digestive processes in the stomach and intestines; we can follow the transport of oxygen, food, and waste products by the blood; the purification activities of the kidneys and the liver are pretty well detailed; the glandular secretion of hormones and their resulting stimulation of specific reactions in remote parts of the body are no longer the mystery they once were.

The Early Chemical
Discoveries

But the physical explanation of the functions of the body organs was not the only important product of early biological research. In the first decades of the nineteenth century the development of the science of chemistry made possible the elimination of some even more profound mysteries of living organisms.

The development of basic understanding of chemical phenomena was initiated by the discovery of the atomic nature of matter. The suggestion in 1803 by the English chemist John Dalton that all matter is made up of tiny, sub-submicroscopic particles led to a shift in the emphasis of the chemists from the bulk characteristics of powders, gases, and solutions to the properties of their constituent building blocks. This more fundamental approach ultimately made of chemistry a science rather than an art. But in the process the science of chemistry turned out to be essentially indistinguishable from the science of physics. The fact that a mixture of hydrogen and oxygen gases combines to form water, when heated to a certain temperature, may appear at first glance to be a purely chemical observation, far removed from the considerations of energy, fields, and forces of the physical scientist. But the difference disappears when the chemical process is analyzed more deeply and found to be only a manifestation of the effects of electric attraction and repulsion among the positively charged nuclei of hydrogen and oxygen atoms and their surrounding negatively charged electron clouds. Even temperature is found to be only another term for the average speed with which the randomly moving gaseous particles bump into one another, thereby influencing the rate of the resulting "chemical" reaction.

In a discussion such as this one the essential reducibility of chemistry to physics is of great importance. Although it remains to be seen whether biology is similarly reducible, such a possibility could not even be seriously considered if it were still thought, as it once was, that chemistry and physics are distinct fields of science with separate governing principles.

But we have digressed. Let us return to our specific biological interest in the development of the physical science of chemistry: it permitted the nineteenth-century investigators to disprove a long-standing idea that the so-called organic chemical compounds appearing in plants and animals are so different from the so-called inorganic compounds of inanimate material that the gap between them can never be bridged by physical methods.

The initial discovery was made in 1827 by a German chemist, Friedrich Wöhler, who showed that the animal waste product urea could be synthesized from the common inorganic chemicals ammonia and cyanic acid. Startling though this discovery was to the scientists of the day, it was soon found rather easy to make many such transformations of matter from inorganic to organic and back again. Ultimately the explanation was provided by the new understanding of the physical nature of chemical phenomena. For every chemical compound was found to consist of building blocks, or *molecules,* each composed of a characteristic three-dimensional configuration of atoms. And exactly the same few dozen different kinds of atoms—hydrogen, oxygen, carbon, and so on—were found to enter into the molecular composition of inorganic and organic substances. Thus common atomic construction materials were used in all molecular building blocks. If the chemist could mix molecules of different inorganic substances and then blend and rearrange their atomic ingredients by the use of heat, light, or electrical energy, there seemed to be no reason, in principle, why an organic molecular arrangement of the material could not result. Apparently, at least in some cases, it did. Well before the end of the nineteenth century these discoveries had destroyed the convenient doctrine that the special properties of living organisms were somehow attributable to a mysterious and unbridgeable gulf forever separating all substances of organic origin from those of inorganic origin.

But the apparent compositional kinship of organic and inorganic materials could not invalidate the observation that great differences did exist between the properties of living and nonliving organizations of matter. After all, only living organisms were able to grow, adapt to environment, and reproduce. These remarkable capabilities certainly did not appear when ordinary inorganic chemicals were mixed together. There was still a "life principle" somewhere.

The nineteenth-century chemists did not find the life principle. However, their discovery that inorganic and organic compounds contain the same atomic ingredients led them to a productive line of investigation. For this caused them to look more closely at the arrangement of the atoms within the molecule, as a possible source of the peculiar properties of living matter. And when the chemical substances which seemed to contribute to living cells their unique properties were isolated and analyzed, they were indeed found to be characterized by an unusual structure: their molecules were enormously large and complex. The molecule of a typical inorganic substance contains only one or two atoms of each of the several elements which enter into its composition. The molecules of such common organic substances as carbohydrates, fats, and oils are somewhat larger, but they rarely contain more than a few dozen atoms. However, the components of living cells that seemed to be most lifelike in their properties were found to have molecules containing thousands or tens of thousands of atoms!

Giant Molecules and Genetic Mechanisms

Many of the exciting biological discoveries of recent years have come from the study of these giant molecules.* We know today

---

* "Giant" is of course a relative term. Even the largest of these molecules is too small to be seen under a high-powered optical microscope.

that they appear in large numbers in every living cell—whether it be that of a bacterium, amoeba, dandelion, or man. It has been learned that there are two general classes of giant molecules, featuring different basic schemes of atomic architecture and therefore different kinds of chemical properties. The first-discovered of the two classes of substance was given the name *protein,* from the Greek, meaning "of first importance." The second class of substance was called *nucleic acid,* in recognition of its acidic properties and of the tendency for many of its most interesting activities to occur in the nuclei of the cells.

We now know a great deal about proteins and nucleic acids. One of the most important early discoveries was that in addition to structural proteins which appear in the various tissues and organs of the body, there are some forms of protein molecules which are *catalysts*—substances that enormously accelerate specific chemical reactions in the protoplasm of the cells. These protein catalysts, which are also called *enzymes,* achieve their results by means of electrical forces of attraction and repulsion operating between the atoms of their molecules and those of the other molecules with which they interact. Through such forces a particular kind of protein molecule may, on contact with an organic "food" molecule floating about in the cellular fluid, cause that molecule to break apart into smaller pieces. Indeed, this is the way our digestive enzymes work. Their remarkable efficiency permits us to break down, or digest, our food far more rapidly than would be the case if our stomachs were filled with the strongest possible concentration of acid—and without the destructive effects on the stomach lining that the acid would have.

There are also constructive protein enzymes, which greatly accelerate the chemical combination of some of the ingredients of the protoplasm. For example, once the digestive enzymes have broken down our food into smaller structural units, other enzymes then force these pieces together to form the different, specifically human kinds of large molecules that our bodies need.

Each protein enzyme is remarkably specific, causing one and only one kind of molecular combination or disruption. But every cell, whether it be a single-celled organism or one of the billions of cells in a human body, contains several thousand different kinds of enzyme molecules. The resulting complex battery of chemical reactions among the protoplasmic ingredients constitutes the *metabolism* of the living cell. Through such metabolism food is digested, waste matter is produced, sensitivity to environmental conditions is determined, membranes and solid cellular structures are formed, and growth occurs. The detailed nature of these lifelike activities is different in the cell of an amoeba from that in the cell of a human being because many of the thousands of kinds of protein enzymes in the amoeba possess numbers and arrangements of atoms in their molecules that are different from those in the human being; this results in different specific chemical reactions and therefore in different kinds of animals.

Their tight control of the complex metabolic processes that contribute to cells most of the properties we call lifelike would seem to entitle proteins to be called "of first importance." However, it turns out that the nucleic acids must be ranked even higher. For the nucleic acid molecules are able to select ingredients from the cellular fluids and put them together to form protein molecules: the nucleic acids manufacture the enzymes. Again, this is a very special process: the detailed atomic structure of the molecule of nucleic acid determines its electrical attractive and repulsive properties and therefore the detailed atomic structure of the final protein product.

The use of a special molecule of nucleic acid as a tool for the manufacture of each protein enzyme molecule might appear to be a pointless complication. However, as is generally true of natural processes, there is logic in the arrangement. It arises out of a second important property of the nucleic acid molecules: they are able to reproduce themselves. For when the chemical conditions are favorable (as determined by the activities of the protein enzymes), the nucleic acid molecules extract from the cellular fluids, not the components of protein molecules, but those suitable for the formation of new molecules of nucleic acid. And the electrical properties of the nucleic acid work in such a way as to cause each new molecule so produced to be an exact, atom-for-atom, copy of the original, even though the generating molecule may contain 100,000 or more atoms in its complex structure!

One final detail completes the logic of the nucleic acid/protein enzyme relationship: from the time that a new cell is formed until growth causes it to split into two cells, self-reproduction results in enough of each kind of nucleic acid molecule to supply the needs of both cells; thus at birth each new cell receives a complete complement of all-important nucleic acid molecules from its parent cell and is therefore able subsequently to generate the specific enzymes needed to give itself the appropriate metabolic pattern; this is why amoebas produce only amoebas and human cells only more human cells.*

In short, it has been found that the protein enzymes are responsible for the complex pattern of chemical reactions that gives each cell its distinctive characteristics, and that the nucleic acids, through self-reproduction and protein manufacture, are responsible for ensuring that each new generation of cells possesses the battery of enzyme substances peculiar to its species. Thus the nucleic acid/protein enzyme mechanisms truly constitute the genesis of the properties of the cells they serve. They are quite properly referred to as the *genetic mechanisms*.

Experimental Confirmation of the Theory

Although extensive detail has no place in this kind of semi-popular treatment, at least a slight reference to the experimental

---

* In the original fertilized egg cell that starts this process for a complex organism such as a human being, half of the generating nucleic acid molecules come from the female sex cell, half from the male sex cell.

support for the conclusions may be in order to help provide some sense of their probable validity. A good test of a biological theory is whether it accurately accounts for the results obtained when the scientist interferes in some calculated way with the normal processes of nature. Let us digress to consider the results of one such series of tests.

The series we shall consider starts with work initiated in 1961 by M. W. Nirenberg and J. H. Matthaei, government research scientists of the National Institutes of Health (NIH) at Bethesda, Maryland. At that time the understanding of the physical mechanisms underlying the relationship between protein and nucleic acid molecules was well enough advanced to suggest that these mechanisms are probably very similar in all organisms. Nirenberg and Matthaei set out to test this conclusion: they developed experimental techniques whereby they could combine nucleic acid molecules from one kind of organism with protoplasmic ingredients from another kind of organism. They reasoned that if their understanding was sound, protein molecules would be generated by this hybrid combination. They were!

Later an experiment by James Bonner and coworkers at the California Institute of Technology provided a logical extension of the results of the NIH research. For, combining nucleic acid from pea plants with protoplasmic components from a common species of bacteria, they not only confirmed the earlier discovery that a hybrid mixture would produce protein but were able to establish that the protein molecules produced, although assembled from ingredients provided by bacterial fluids, possessed the properties of protein molecules naturally occurring in pea plants.

Meanwhile, Nirenberg and Matthaei were able to make a spectacular extension of their own original results. They did so by replacing the natural nucleic acid of their first experiment with a synthetic, man-made product. Chemical techniques were not yet far enough advanced to permit the synthesis of nucleic acid molecules of the size and complexity of most natural ones, but simple forms could be put together out of the ordinary inert ingredients available in the chemical laboratory. This was done, and the original experiment was repeated. And the synthetic nucleic acid manufactured protein molecules—simple ones, corresponding to the simple structure of the generating nucleic acid molecules, but real protein nonetheless. Creation of life in the test tube? Depending on the definition of life—perhaps!

Indeed, investigators employing experimental techniques similar to those of Nirenberg and Matthaei have succeeded in deciphering most of the so-called genetic code. By determining the specific atomic structure of the protein molecules resulting from the manufacturing activities of synthetic nucleic acid of known structure, they have been able to work out many of the details of how a nucleic acid molecule exerts precise architectural control over the construction of a protein molecule.

The Major Conclusion

So much for our short side trip into the experimental laboratory. For our purposes, the important observation about the large amount of this kind of modern research is that it is steadily strengthening the already convincing case for a simple clear-cut conclusion: the properties we call lifelike that differentiate living cells from inanimate matter result directly from the nucleic acid/ protein enzyme mechanisms, and a complete explanation of these mechanisms is possible in terms of the operation in inert chemical ingredients of the ordinary laws of physical science.

BIBLIOGRAPHY

Asimov, I., *The Genetic Code* (The Orion Press, Inc., New York, 1962), chap. 4, "The Building Blocks of Protein"; chap. 5, "The Pattern of Protein"; chap. 6, "Locating the Code"; chap. 7, "The Cinderella Compound"; chap. 8, "From Chain to Helix"; chap. 9, "The Cooperating Strands"; and chap. 11, "Breaking the Code."
Asimov, I., *The Intelligent Man's Guide to Science* (Basic Books, Inc.,

Publishers, New York, 1960), chap. 10, "The Molecule," and chap. 11, "The Proteins."

Asimov, I., *The Wellsprings of Life* (Abelard-Schuman, Limited, New York, 1960), chap. 11, "Building Blocks in Common"; chap. 12, "The Shape of the Unseen"; and chap. 13, "The Surface Influence."

Beadle, G. W., "The New Genetics," in 1964 Britannica *Book of the Year* (Encyclopaedia Britannica, Inc., Chicago, 1964).

Bonner, J., R. C. Huang, and R. V. Gilden, "Chromosomally Directed Protein Synthesis," *Proceedings of the National Academy of Sciences,* vol. 50 (November, 1963), pp. 893–900.

Moore, R., *The Coil of Life* (Alfred A. Knopf, Inc., New York, 1961), chap. 4, "Wohler and Liebig: Makable by Man," and chap. 17, "Pauling and Sanger: The Proteins—Another Coil."

Morowitz, H. J., *Life and the Physical Sciences* (Holt, Rinehart and Winston, Inc., New York, 1963), chap. 2, "Atoms and Cells," and chap. 4, "Large Molecules."

Nirenberg, M. W., "The Genetic Code: II," *Scientific American,* March, 1963, pp. 80–94.

Wooldridge, D. E., *The Machinery of Life* (McGraw-Hill Book Company, New York, 1966), chap. 2, "The Basic Construction Materials"; chap. 3, "The Biological Building Blocks: The Simpler Organic Molecules"; chap. 4, "The Building Blocks 'of First Importance': Protein Molecules"; chap. 10, "The Generation and Reproduction of Nucleic Acid Molecules"; chap. 11, "The Development of Architectural Talents by the Nucleic Acids"; chap. 12, "Refinement of the Nucleic Acid Mechanisms"; and chap. 13, "The Proliferation of Successful Techniques: Biological Standardization."

chapter 2

The Origin of Living Cells

From a review of the chemistry of living organisms we have learned that the lifelike properties of cells appear explicable in terms of the ordinary laws of physics. However, nonphysical principles might still be required to explain how life developed: the susceptibility of the *properties* of currently living organisms to a purely physical explanation does not necessarily imply that only ordinary physical principles were involved in their *origin*. If the first American astronauts to set foot on the moon were to find there a message in Russian scratched on its surface, their ability to hypothesize a series of meteoritic impacts capable of producing just the observed configuration of scratches would hardly be likely to convince them that it really happened that way. And in our case, are we to believe that the blind workings of the ordinary laws of physics, without the intervention of some vitalistic guiding principle, can account for the original appearance of the precisely specified, fantastically complicated nucleic acid molecules on which modern life depends?

Science has not ignored this important question. Recent research has thrown light on the probable origin of life, as well as on its present properties.

The Origin of
Nucleic Acid

The paleogeologist and astrophysicist have participated with the biologist in research on the origin of life on earth. Their knowledge has made it possible to specify the probable composition of the atmosphere that surrounded the primordial earth shortly after it was formed, about five billion years ago. This, in turn, has permitted experiments with gaseous mixtures simulating the early atmosphere. In particular, such mixtures have been subjected to the forms of energy that are believed to have abounded in early days—heat, lightning discharge, ultraviolet radiation, and radioactive bombardment. The rather spectacular outcome of these fundamentally simple experiments has been the discovery that such treatment automatically results in the formation of many simple varieties of organic chemical products. And if these products continue to be exposed to energy bombardment while they are allowed to settle into warm water simulating the seas that covered most of the primordial earth, chemical combinations are found to occur in such a way as to form more complex substances, including even molecules of nucleic acid!

To be sure, while complex relative to the starting atmospheric ingredients of the experiment, the nucleic acid molecules formed in this way are still small and simple compared with the giant molecules of today's organisms. To understand what this difference in complexity means, we need to know that the molecule of a nucleic acid has a beadlike structure, each bead, or *nucleotide,* consisting of a three-dimensional arrangement of thirty-one to thirty-six atoms of carbon, hydrogen, oxygen, nitrogen, and phosphorus. In all, eight different types of nucleotide appear as building blocks in the many thousands of different kinds of nucleic acid molecules,* each nucleotide being characterized by a con-

---

* Only four types of nucleotide appear in any one molecule. But there are two slightly different subclasses of nucleic acid, called *ribonucleic*

figuration of its approximately three dozen atoms that is slightly different from that of the other seven types of nucleotide. And one kind of nucleic acid manufactures a different kind of protein than another kind of nucleic acid manufactures only because the specific sequence of nucleotides strung together to form its long molecule is different from the sequence of the other molecule. The nucleic acids that play such a vital role in modern organisms achieve their remarkable properties through a complexity of structure characterized by hundreds or thousands of nucleotides in each molecule. By comparison, the molecules resulting from laboratory experiments on simulated primordial atmospheres rarely contain more than a half-dozen nucleotide segments. Physical science, to be adjudged adequate to account for the origin of life, must provide a connection between the simple organic molecules created by energetic processes in the primordial atmosphere and the complex molecules of modern organisms.

Fortunately, there is an explanation of how complexity may have developed. It depends for its plausibility on the many millions of years that were available for natural processes to modify the original simple molecules. The construction of such an explanation from the slender clues available so long after the fact poses a challenging problem to the modern scientist, and we shall not be able to go into all the details here. Nevertheless, a review of some of the principal features of the reconstruction of the prehistory of life is possible, and may serve to provide some feeling for the probable validity of the resulting conclusions.

Droplets: Precursors of Living Cells

One of the requirements for the development of complex organic substances was some mechanism for bringing together and

---

acid (RNA) and *deoxyribonucleic acid* (DNA). The four nucleotide types making up RNA molecules are slightly different from those making up DNA molecules.

concentrating the essential ingredients to facilitate their chemical interaction. This was automatically provided by certain electrical cohesive effects, which cause suspensions of large molecules in water to aggregate into droplets. The spherical surfaces of such droplets, because of the electrical properties of the contents, also frequently possess special permeability features. For instance, small molecules may be able to pass easily into or out of a droplet, while large molecules are trapped and held.

Now the relatively large organic molecules—carbohydrates, oils, nucleic acids, and the like—that aggregated to form the first droplets in the primordial seas orginated among the ingredients that rained out of the skies and were formed as a result of energetic processes in the atmosphere similar to those recently duplicated in the experimental laboratory. But most of the material formed in the atmosphere, just as in the laboratory experiments, must have consisted of pieces of organic molecules rather than of complete molecules. Thus the amount of high-molecular-weight material available for the formation of droplets was only a tiny fraction of the partially processed ingredients, which were too small to be confined by the droplet surfaces. Conditions were right for some kind of "technological invention" to increase the efficiency of droplet formation.

Such an "invention" was inevitable. It consisted in the accidental linking together of several molecular fragments to form a structure possessing special catalytic effectiveness—the ability to promote the combination of other of the available small-molecular ingredients to form complex molecules.* Such a catalytic molecule, because of its own relatively large size, would tend to be trapped within a droplet. This would make of the droplet a

---

* We have already encountered such catalytic action, in our introduction to the protein enzymes. Of course, nothing like the enormous catalytic effectiveness of modern enzymes could have been attained by simple, randomly formed molecules. However, many simple structures are known today that, given the centuries or millennia available in this primitive period of development, could have led to profound changes in the chemical content of the early seas.

sort of chemical factory, in which some of the small-molecular ingredients seeping in from the surrounding sea water would be combined into more complex substances, which would in turn be trapped by their own large-molecular structure. Thus the droplet would grow, by the steady accumulation of the trapped materials.

Of course, the use of the singular in the narrative is only a literary device. In the almost numberless random collisions occurring among the molecular fragments of the primordial seas during the millions of years we can allow for this part of the prehistory of life, the workings of the physical laws of probability must have resulted in the generation of very large numbers of effective catalytic molecules. And many substantially different kinds of catalysts would have been produced in this way. In particular, sooner or later the phenomenon known today as *autocatalysis* would have appeared. In autocatalysis one of the products of the reaction stimulated by the catalyst is more of the catalyst itself.

A single droplet possessing this kind of chemical factory could have led to spectacular changes in the chemical content of the early seas. For such a droplet would have been able not only to grow but also to reproduce its own kind. Ultimately, by the operation of the normal physical laws related to surface tension, volume, and weight, the growing droplet would have broken up into smaller parts. But since autocatalysis would have provided the original droplet with many catalytic molecules instead of just one, some of the smaller, "baby" droplets into which the "parent" fragmented would likely have contained some of the all-important catalytic material, causing them to embark on the same kind of chemical growth activities that had previously been the specialty of the parent.

By such developments the droplets that started as inert bags of chemicals could slowly lead to structures with properties of growth, metabolism, and reproduction at least crudely similar to those exhibited by modern single-celled organisms.

Evolution

Once the first autocatalytic droplets appeared, they must have quickly given rise to a major population explosion. For each baby droplet would ultimately have reproduced more of its own kind when growth led to its own instability and disintegration. If the second generation consisted of several droplets of the new reproducing type, the third generation may have consisted of a dozen, the fifth of a hundred, the seventh of a thousand, the thirteenth of a million, and so on. A few thousand years of such proliferation could well have resulted in sweeping the seas of most of their organic fragments and incorporating them in the form of large molecules in these curious new and almost lifelike organizations of matter.

It is not reasonable, of course, to assume that only one type of autocatalytic droplet developed and prospered. Various self-aggrandizing chemical cycles were physically possible, and were doubtless discovered by nature's random processes. Droplets of differing detailed chemical metabolism must therefore have competed for the limited supply of organic molecular fragments in the sea on which growth depended. And some metabolic cycles must have been more efficient than others. Those droplets which luckily possessed a set of ingredients contributing to fast metabolism (rapid chemical processing) would have ingested a proportionately large fraction of the available "food." Therefore they would have come to maturity more quickly and had more progeny than other, less well-endowed species of droplets. Gradually the rapidly proliferating species would have starved the others out of existence.

But just as the random jostling of molecular fragments resulted in the first place in the formation of catalytic molecules, accidental changes in their structure, or *mutations,* must subsequently have occurred from time to time. Once in a long while one of these changes would have resulted in a molecular form permitting a new cycle of chemical reactions which materially increased the growth rate of the droplet. The resulting new species, through its superior rate of proliferation, would ultimately have starved out and displaced the old-fashioned kinds of droplets. There is a name that is commonly applied to such a competition for a single source of raw materials among a number of self-aggrandizing chemical systems each of which is subject to small random changes affecting its rate of activity—it is called *evolution*. We are here dealing with evolution at a more primitive level than that envisioned by Darwin, to be sure, but it is essentially the same phenomenon. Though generally considered a biological principle, evolution, like all the other natural processes we shall consider, is revealed on analysis to be a straightforward consequence of the operation of the ordinary laws of physics and chemistry.

Evolution of the Nucleic Acids

The species-improvement capabilities of evolution, when operating on the droplets for millions of years, must ultimately have led to remarkably increased sophistication in the metabolism of these precursors of living cells. Complex cycles of chemical reactions, sustained by a number of enabling catalysts, would automatically have appeared because of the accelerated growth rates and improved survivability features they provided. With the development of such complex metabolism, the conditions were finally right for the nucleic acids, which have been missing from the recent narrative, once again to take the center of the stage.

For the remarkable architectural talents of nucleic acid molecules, which are so fundamental to the metabolism of all modern organisms, require for their exercise considerable assistance from the surrounding environment. Even though the small-molecular constituents of protein enzymes were undoubtedly available among the early atmospherically supplied ingredients, their assembly into large molecules by nucleic acid requires the presence of certain special catalysts and other chemical substances that could hardly have appeared in the early simple droplets. Also unlikely would have been the early appearance of the different but equally complex chemical environment that sometimes permits a nucleic acid molecule to control the selection and precise assembly of suitable ingredients to form identical copies of itself.

But as evolution of the droplets provided them with increasing numbers of complex catalysts and other new forms of organic molecules, a time must have come when mutations could occasionally result in new species possessing internal chemical conditions suitable for the support of the special functions of the nucleic acid molecules. In such droplets self-reproduction would have provided a high concentration of these molecules, while their other unique architectural talent would have resulted in an abundant supply of protein molecules. But we have seen that proteins can be unusually effective catalysts. To be sure, the generating nucleic acid molecules, and therefore the resulting proteins, were in the period in question simple by comparison with today's giant and remarkably effective structures. Nevertheless, once in a thousand or million times that accidental changes in droplet chemistry permitted protein manufacture to occur, the specific molecular structure of the nucleic acid that happened to be in the droplet might have resulted in a structure of the derived protein providing catalytic properties contributing some increase in growth rate to the host droplet. In the ensuing years, the routine operation of the principles of evolution would then have brought this new species to ascendancy, as it starved out its less effective competitors.

Evidently the catalytic effectiveness of the early proteinlike molecular structures was substantial, for eventually the only surviving species of droplets were those equipped with nucleic acid molecules and a chemical metabolism permitting their self-reproduction and protein manufacture. Although initially the molecules of the controlling nucleic acid and therefore those of the resulting proteins must have been small and simple by present-day standards, with the passage of time there would have been a tendency toward greater complexity. The usual principles of evolution would have led to this result: Accidental changes would have occasionally modified one of the molecules of nucleic acid in a droplet. One of the simplest and therefore statistically most likely changes would have been the linking of a new nucleotide segment to an existing molecule. If the larger protein manufactured by the new kind of nucleic acid was more effective in catalyzing the growth processes, the enhanced proliferation of droplets of the new design would ultimately have led to their numerical superiority. Since the larger protein catalysts, which are produced by the larger nucleic acid molecules, are indeed usually more effective, there would have been a steady tendency toward nucleic acid complexity in the evolutionary development of successive species of what we must now refer to as living cells.

The Major Conclusion

It is not easy for scientists to reconstruct the history of chemical events that took place long before the deposit of the fossil records upon which our relatively clear view of the last billion years of biological development is based. No one pretends that such a sequence as that just outlined is a completely true description of the past. All that is claimed is that it is probably "true to life" in that the events that it portrays are similar enough in quality to those that actually transpired to lead to generally valid conclusions about the nature, although not necessarily the details, of the prehistory of biology.

The important conclusion is that it is possible to put together a fairly convincing story describing how modern living cells could have developed. The story satisfactorily accounts for the origin and subsequent increase in complexity of nucleic acids to the giant molecules of modern organisms. The principles of organic evolution constitute the thread that holds the narrative together. And the principles of evolution, as we have seen, are accounted for by the laws of physics. In short, the prevailing line of research and theory provides for the origin as well as for the chemistry of life an explanation based entirely on the inexorable workings of the ordinary laws of physical science in the inert ingredients of the earth.

BIBLIOGRAPHY

Asimov, I., *The Intelligent Man's Guide to Science* (Basic Books, Inc., Publishers, New York, 1960), chap. 11, "The Proteins."

Asimov, I., *The Wellsprings of Life* (Abelard-Schuman, Limited, New York, 1960), chap. 13, "The Surface Influence."

Beck, W. S., *Modern Science and the Nature of Life* (Doubleday & Company, Inc., Garden City, N.Y., 1957), chap. 6, "The Unit of Life."

Calvin, M., *Chemical Evolution* (University of Oregon Press, Eugene, Ore., 1961).

Ehrensvard, G., *Life: Origin and Development* (The University of Chicago Press, Chicago, 1960), chap. 7, "Out of Dust and Fire," and chap. 10, "Activity."

Jackson, F., and P. Moore, *Life in the Universe* (W. W. Norton & Company, Inc., New York, 1962), chap. 2, "The Nature and Origin of Living Organisms on the Earth."

Oparin, A. I., *The Chemical Origin of Life* (Charles C Thomas, Publisher, Springfield, Ill., 1964), chap. 1, "Initial Stages in the Evolution of Carbon Compounds"; chap. 2, "The Formation of the Primaeval Broth"; chap. 3, "The Origin of the Earliest Organisms"; and chap. 4, "The Further Evolution of the Earliest Organisms."

Oparin, A. I., *Life: Its Nature, Origin and Development* (Academic Press Inc., New York, 1962), chap. 2, "The Origin of Life."

"The Origin of Life," special report in 1965 Britannica *Book of the Year* (Encyclopaedia Britannica, Inc., Chicago, 1965).

Wooldridge, D. E., *The Machinery of Life* (McGraw-Hill Book Company, New York, 1966), chap. 5, "The Paleontology of Protein"; chap. 6, "The Primeval Lakes: Giant Precursors of Living Organisms"; and chap. 7, "Miniaturization of the Primordial Pools: Progress toward Living Cells."

chapter 3

Multicellular Organisms

Despite an occasional reference to human cells and physiology, the story told to this point pertains primarily to very simple forms of life. We must now move along to a consideration of the origin and properties of complex organisms. In this and the next chapter we shall examine the reasons why essentially all scientists believe the development of modern plants and animals to be understandable in terms of the continued operation of the same physical principles of organic evolution that produced the first single-celled forms of life.

Two Problems

Credibility of the theory of the natural evolutionary origin of multicellular organisms rests on solutions of two interrelated and long-standing problems. The first problem may be simply stated: By self-reproduction, the nucleic acid molecules in a growing cell multiply so that at cell division, each of the resulting two cells is supplied with equal and adequate quantities of the same kinds of molecules. But through their manufacture of

29

protein enzymes, the particular kinds of nucleic acid molecules in the cell determine the particular kinds of chemical reactions that constitute its metabolism and therefore specify what kind of a cell it is. Thus amoebas split into more amoebas, human cells into more human cells. But doesn't this also mean that kidney cells could lead only to more kidney cells, and heart cells to more heart cells? And if so, when a plant or animal develops from a single fertilized egg cell, how can its different parts ever have different properties?

The interrelated second problem is equally easy to formulate: If the nucleic acid molecules are indeed all-powerful in determining the properties of their host cells, does this not require that the original sperm and egg cell from male and female parent contain enough different kinds of nucleic acid molecules to specify the literally hundreds of thousands of different chemical reactions that occur in the various parts of a complex plant or animal? Is it reasonable that such a complete "book of instructions" for the assembly and operation of a human being, say, should all be contained in the tiny fertilized egg cell from which the individual develops? And if the original cell does indeed contain such a book of instructions for the entire organism, do each of the cells into which it splits and each of the cells into which these new cells split, and so on, contain a similarly vast amount of manufacturing information?

Completeness of the Book of Instructions in Each Cell

The second of the two problems was solved first. Indeed, evidence bearing on the completeness of the book of instructions in each cell has been available for many years. As early as 1891 the German biologist Hans Driesch threw some light on this important question. He employed the common sea urchin in his experiments. Like all other higher animals, the sea urchin begins its development with a series of cell divisions which are quite precise. First the original fertilized egg divides into two cells, then four, then eight, then sixteen, and so on. Driesch discovered that if the individual cells are separated from one another by violent shaking in the four-cell stage, each of the resulting separated cells will go on to develop into a complete sea urchin. Later experiments demonstrated the same principle with other animals, including vertebrates. In fact, the occurrence of identical human twins is ascribed to some form of intrauterine event that separates into two embryos the two cells that develop from the original fertilized egg.

Recent work by J. B. Gurdon, zoologist at Oxford University, has provided even more convincing evidence that every cell contains a complete book of instructions for the entire organism. For he found that the original egg cell of a species of African aquatic frog would develop into a complete and normal animal even after its nucleus (containing the nucleic acid) had been replaced by one removed from a specialized intestinal cell of a fully developed tadpole of the same species.

The examples cited are not unique. Much evidence is now available showing that every cell of a complex plant or animal contains a set of nucleic acid molecules complete enough to specify the chemistry, and therefore the structure, of every part of the entire organism. Thus if we ever succeed in completely translating all aspects of the genetic code, we should be able to deduce all the inherited characteristics of an individual—color of the eyes, shape of the nose, contours of the face, pattern of the hairline—from analysis of the nucleic acid from a single heart cell, nerve cell, skin cell, or liver cell!

This conclusion is encouragingly consistent with our general understanding of the genetic mechanisms. We must now consider the other problem: how can different types of cell coexist in the same plant or animal? The answer has come only from very recent research. Indeed, the study of the mechanisms underlying the phenomenon of cell differentiation is one of the most active lines of current biological investigation.

The Sources of Cell Diversity in Complex Organisms

Two important processes have been discovered that contribute to the achievement of diversity by the cells of a complex organism. One process consists in a modification of the effectiveness of the protein enzymes by some of the contents of the cellular fluids. Molecules of certain substances can wrap themselves around enzyme molecules and inactivate them, while the attachment of molecules of other substances can increase the catalytic effectiveness of the enzymes. Therefore, if cells in different regions of a multicellular organism are washed by fluids of differing chemical content, some of their enzymatically catalyzed reactions may proceed at different rates, leading to differences in the chemistry and structure of the cells.

The other process that is now known to contribute to cell differentiation is one which acts directly on the nucleic acid molecules, rather than on the resulting enzymes. Chemicals in the cellular fluids can "turn off" or "turn on" the protein-manufacturing capability of particular types of nucleic acid molecules and thereby selectively decrease or increase the supply of the various types of catalysts. This phenomenon has been given the name *gene switching,* inasmuch as the nucleic acid molecules that are activated or inactivated in this way are frequently called *genes.*

Gene switching can have very extensive effects on the properties of cells and organisms. For example, one investigator cultured cells from the nervous system of an amphibian until they developed properties identifying them as early pigment cells with a characteristic content of color granules and a star-shaped contour. He then removed the cells from the culture medium, washed them, and placed them in a different medium. As a result some of the cells, without undergoing division, appeared to change their properties entirely, going so far as to develop muscle fibrils with a recognizable cross-striation, as though they had changed their type from nerve to muscle cells.

Even more noteworthy is the curious adaptability of certain uni-cellular organisms. *Naegleria,* for example, move around in an amoeboid form when they are on a relatively dry substratum in the presence of bacteria; but when they are surrounded by more or less pure water, they change their structure entirely and develop flagella at one end, with the whole body elongated into a characteristic flagellate type. And a much more complex organism, the Mexican axolotl, which normally lives all its life and reproduces itself as a gilled newt in the water, can be turned into a land salamander by a single dose of thyroid. Despite the fact that the axolotl has lived its life aquatically for thousands of generations, a fraction of a milligram of thyroxin, even from a sheep or fish, will bring out the latent salamander in a couple of weeks.

Indeed, the induction of major changes in an entire plant or animal by the influx into the cells of gene-switching chemicals is so common that it has been observed by all of us—in the metamorphosis of a tadpole into a frog or the pupation of an insect, for example. These changes can be induced at any time by dosing the young tadpole or larva with suitable hormones. Familiarity may blind us to the truly spectacular nature of such transformations. The frog differs so much from the tadpole, and the butterfly from the larva, as to require that each cell, in its complement of nucleic acid molecules, carry specifications for essentially two different species of organism, with switching arrangements to turn off one set of controlling molecules and turn on the other upon receipt of the proper hormonal signal. In the face of such natural phenomena, one wonders whether the fairy-tale conversion of Cinderella's white mice into footmen was so farfetched after all!

The Major Conclusion

The chemical modification of the catalytic effectiveness of the enzymes and the even more powerful phenomenon of gene switching can cause different groups of cells in the same organism, despite a common genetic heritage, to develop widely different structural and metabolic properties. In principle, therefore, all that is necessary to account for cell diversity is an explanation of how natural causes result in the imposition of different chemical environments on different parts of the organism. The embryologists have been struggling with this problem for over a hundred years. In the next chapter we shall learn about some of their interesting discoveries.

BIBLIOGRAPHY

Barry, J. M., *Molecular Biology: Genes and the Chemical Control of Living Cells* (Prentice-Hall, Inc., Englewood Cliffs, N.J., 1964), chap. 6, "How Genes Control the Formation of Other Cell Molecules."

Changeux, J. P., "The Control of Biochemical Reactions," *Scientific American,* April, 1965, pp. 36–45.

"Experimental Embryology," *Encyclopaedia Britannica,* 1962 ed.

Gurdon, J. B., "The Transplantation of Living Cell Nuclei," *Advances in Morphogenesis,* vol. 4 (1964), pp. 1–42.

Huang, R. C., and J. Bonner, "Histone: A Suppressor of Chromosomal RNA Synthesis," *Proceedings of the National Academy of Sciences,* vol. 48 (July, 1962), pp. 1216–1222.

Oparin, A. I., *The Chemical Origin of Life* (Charles C Thomas, Publisher, Springfield, Ill., 1964), chap. 4, "The Further Evolution of the Earliest Organisms."

Vogel, H. J., "Control by Repression," in *Control Mechanisms in Cellular Processes,* ed. by D. M. Bonner (The Ronald Press Company, New York, 1961), pp. 23–65.

Waddington, C. H., *New Patterns in Genetics and Development* (Columbia University Press, New York, 1962), chap. 4, "Morphogenesis in Single Cells."

Wells, H. G., J. S. Huxley, and G. P. Wells, *The Science of Life* (Doubleday & Company, Inc., Garden City, N.Y., 1938), book 4, chap. 4, "The ABC of Genetics," and book 4, chap. 5, "The Growth of the Individual."

Wooldridge, D. E., *The Machinery of Life* (McGraw-Hill Book Company, New York, 1966), chap. 15, "Genetic Mechanisms and Cell Diversity in Multicellular Organisms."

Zalokar, M., "Ribonucleic Acid and the Control of Cellular Processes," in *Control Mechanisms in Cellular Processes,* ed. by D. M. Bonner (The Ronald Press Company, New York, 1961), pp. 87–140.

chapter 4

Dandelions and Men

We have developed essentially the following picture of how cell differentiation occurs in plants and animals: As the fertilized egg divides—first into two cells, then into four, then eight, and so on —slight differences in the chemical conditions in the several cells cause corresponding slight differences to occur in their internal reactions. These differences affect the chemical composition of the fluid emanating from the cells and therefore supplement the already existing sources of diversity to cause still later cells to be even more varied in their properties. As the organism grows by further cell division, the compounding of effects ultimately results in the appearance of cells of widely differing structure and metabolism. The mechanisms that respond to the local chemistry to yield different types of cells were discussed in the preceding chapter. Let us now consider the origin of the chemical differences that initially set the early cells on divergent paths of development, as well as some other natural effects that magnify the local environmental differences as the embryo grows.

Differentiating Agents in
the Fertilized Egg

A good way to start these considerations is to resolve an apparent inconsistency between our present thesis and experimental evidence we considered earlier. Specifically, we need to reconcile Driesch's discovery that any of the early cells of a sea-urchin embryo can develop into an entire animal with our present notion that differences must begin to appear in the structure and metabolism of the first few cells of a developing organism.

A Swedish embryologist, Sven Hoerstadius, resolved the apparent discrepancy. In a series of beautifully planned and executed transplantation and isolation experiments he was able to show that the material of the sea-urchin egg is not uniform throughout in its composition, but contains varying proportions of two different chemical agents that have important influences on cellular metabolism. One of these agents was found to be concentrated at one pole of the fertilized egg, the other at the opposite pole. Hoerstadius established that it was only because the early cell divisions involved cleavages parallel to the polar axis that the first few cells all possessed the same proportions of the two agents and were therefore able to develop into entire animals. If by artificial means the initial cleavage was forced to occur along such a plane that most of one stimulating agent went into one cell and most of the other into the second cell an entirely different result was obtained: when these two cells were separated, each developed into only part of an animal, and the particular organs which developed in one "half-embryo" were different from those which developed in the other. Evidently one of the two chemical agents discovered by Hoerstadius acts as a switch to turn on the genes responsible for the construction of part of the animal, the other as a switch to turn on those responsible for the construction of the rest. And these two stimulating ingredients are already localized in different regions in the original fertilized egg.

A similar polarity of the original egg cell has been demonstrated for many other species. In some cases at least gravity appears to be the cause of the separation of the different ingredients in the egg cell as it develops in the mother's body. This is true of frogs' eggs, for example. Their polar axes are easily visible because of color differences; in particular, the bulky and nutritive yolk can be seen concentrated near the bottom of the egg. The cells that grow from this half of the egg ultimately develop into the tail parts of the frog; those that grow from the top half become the head parts. Thus the vertical axis of the egg controls a most important structural feature: it determines the body axis of the developing embryo.

While the environmental factor determining the body axis is gravity, quite a different factor determines the plane of symmetry. This plane is determined, for frogs, by the point of entry of the male sperm cell into the egg during the act of fertilization. The meridian of longitude of the sperm entry point becomes the head-to-tail belly line of the animal, while that on the opposite side from the point of entry becomes the line of the backbone.

Structure Determination by Gross Physical Effects

As the embryo develops, a number of gross physical effects come into play in determining, not the metabolism, but the structural arrangement of the various groups of cells. The pressure of water that collects in the growing tissue is instrumental in shaping some of the body cavities. The rolling up of aggregations of certain types of cells to form tubes—such as the *neural tube,* which gives rise to the nervous system—occurs because of electrical attractive properties that cause such cells to try to make as close contact as possible with one another.

Many such physical factors must operate in conjunction with

the specific chemical properties of the cells, to cause different parts of the growing embryo to form different kinds of structures. And such characteristic structural developments in turn can affect the local chemical environment by bringing different types of cells together and thereby exposing each type to the influence of the fluids secreted by the other. Then the cells involved can respond by becoming even more chemically different than before from other cells in the same organism.

Progressive Freezing of Cell Properties

The chemical individuality of the cells usually develops gradually rather than abruptly. If, in a very young embryo, cells are surgically transplanted to the head region from a part of the organism that would ordinarily develop into a tail, the transplanted cells take on the characteristics of the material that surrounds them and grow into component head parts. But if such transplantation is deferred for a time, a fixation of properties occurs so that after the operation the transplanted cells grow into tail-like parts, despite the fact that they must then protrude from the head of the full-grown embryo. It is even possible to follow the gradual development of this fixation of properties: at intermediate stages of growth a small group of cells transplanted from any part of the tail region of the embryo will develop into an entire tail; later cells transplanted from a particular part of this region will develop into only a specific part of the tail. With the passage of time the cells seem to become more and more specialized and therefore more limited in the variety of structures they are able to form.

As might be expected, more rapid cell differentiation sometimes occurs when two previously separated parts of a growing embryo come together. The classical example of this phenomenon is the formation of the crystalline lens in the eye of a vertebrate. By following the gradual development of the embryo (of a chicken, for example), it can be observed that the main body of the eye is formed from brain tissue, by a sort of dimpling or hollowing-out process. This results in an *eye cup,* whose open end gradually grows toward the layer of skin that surrounds the brain parts. When growth finally brings the rim of the eye cup in contact with the surrounding skin, a reaction occurs that causes the circular sector of skin contacted by the cup to embark upon a new process of cellular development. Before long this piece of skin thickens, detaches itself from the surrounding tissue, changes its shape, and becomes incorporated into the eye cup to form the crystalline lens upon which clear vision ultimately depends.

Research has confirmed the validity of the obvious inference: the tissue forming the eye cup in the brain structures contains a chemical which, upon contact with the surrounding skin, triggers metabolic processes that result in the formation of a lens. It has even been shown that the triggering substance is versatile enough to induce the formation of lens structures in other kinds of skin than that of the head region. For example, the eye cup can be transplanted to the flank; upon contact with this kind of skin, a lens is formed. Or skin from another part of the body can be substituted for the head skin in the vicinity of the eye cup; upon contact of the two kinds of tissue, a lens develops. Evidently the stimulating ingredient in the eye cup is able to trigger the genes in the cells of any of the organism's skinlike tissue to reorganize their metabolic processes in the way necessary to develop into a crystalline lens.

The stimulating ingredient is even more versatile; for example, the eye cup of a frog can induce a lens in head or flank epidermis of a salamander embryo.* However, the lens induced has the specific characteristics of a salamander eye rather than a frog eye.

---

* In embryos the immunological reactions that cause transplanted foreign tissue to be rejected do not occur. Hence transplantations are possible not only between different species but between different genera, families, orders, and classes. For instance, mouse tissue can grow in the chick embryo.

Again, when flank skin of a frog embryo is transplanted to the head of a young salamander embryo, the new chemical environment induces the transplanted skin to form head structures. But in doing so it follows its own genetic repertory: it manufactures the horny jaws and teeth characteristic of the frog instead of the dentine teeth characteristic of the salamander. The explanation is, once more, entirely consistent with the picture we have developed of the genetic mechanisms. All cells in the frog contain nucleic acid molecules capable of directing the metabolic reactions involved in the formation of any of the many body structures, and at transplantation the chemistry of the flank-skin cells has not yet become so frozen as to prevent considerable flexibility in their ultimate development. When stimulated by the kind of substance that switches on the genes appropriate for the generation of head-like structures, these flank-skin cells proceed to form such structures. However, whether in the original frog body or in another species after transplantation, the particular structures formed must be controlled in their detailed architecture by the genes of the frog. Hence the transplanted tissue must give rise only to froglike structures, not to salamanderlike structures.

Problem of the Nervous System

An embryological problem of unusual importance and difficulty is posed by the nervous system. How can we account for the enormous mass of specialized nerve cells, or *neurons*—ten billion of them in a single human animal—that seem able to extend their tiny fibers many inches or even a few feet to make highly specific connection with other nerve cells or organs? To be sure, the general notions we have developed as to how cells in the embryo become more and more specialized can cause us to feel that the natural physical and chemical effects we have been dealing with might be adequate to produce even as strange a structure as that of the nerve cell. But how are we to explain the fantastically complex "wiring diagram" that appears to govern the interconnections of so many separate neuronal units?

Because of the difficulty of conceiving of any embryonic growth mechanism that would automatically result in the precise, genetically predetermined point-to-point wiring of such a gigantic set of neuronal components, much attention has been devoted to a simplifying hypothesis—that the neurons in the embryo just grow like Topsy in uncontrolled fashion, making random interconnections as they come in contact with one another. This hypothesis depends for its credibility on the theory that learning processes are subsequently able to strengthen and weaken the various neuronal connections so as ultimately to provide the coordination of physical and mental activity that constitutes the unique accomplishment of the nervous system.

In the higher animals, at least, any general theory of neuronal connectivity must be able to account for the phenomena associated with vision. In the human being, more than half of the several million nerve fibers that connect the brain with other parts of the body go to the eyes. If a random-connection/subsequent-learning theory is the complete answer to our information-handling problem, we would expect that meaningful visual patterns would not occur in newly formed nervous systems, but would arise only after a period of trial-and-error learning. One of the most interesting experiments testing this hypothesis was performed by R. W. Sperry, professor of biology at the California Institute of Technology. In his work he took advantage of the fact that the nervous systems of some animals possess a regenerative capacity: if nerves are cut, they grow back and ultimately function once again. Sperry was therefore able to establish, in an adult toad, a special kind of "newly formed" visual system. He performed an operation wherein the optic nerves were cut and reconnected inversely—that is, the right eye was connected to the nerve from the brain that previously had gone to the left eye, and vice versa. Of course, in such an operation, "reconnecting" consists only of butting the cut ends of the nerves together and waiting for natural processes to reestablish connections from the many tens of thousands of cut fibers to the brain. Even in an uncut optic nerve, these fibers cross and twist in what appears to be a highly random fashion. Such twistings and turnings (which are also characteristic of the human optic nerve) had always contributed to the difficulty of believing that there is any precise built-in pattern of interconnection between retinal receptors and neurons in the brain; this had lent strength to the hypothesis that visual capability is acquired through learning and is not wired in.

Yet, after a few weeks, Sperry's toad was able to see again, apparently as well as ever! In normal toadlike manner it again reacted to the presence of a moving fly by darting out its tongue for the food. This and other observations led to the conclusion that somehow the fibers proceeding from the receptor cells of the eye had managed to seek out and reconnect themselves, one by one, with the neurons of the visual cortex in such a way as to reestablish in the brain a clear image, with normal topological properties of up/down and right/left continuity. That this had nothing to do with learning was proved by an interesting anomaly in the toad's new behavior: if a fly appeared opposite the toad's right eye, it darted its tongue out to the left to attempt to capture it; if the food appeared to the left, the toad would always strike to the right. To the animal, since the optic nerve of the right eye was now connected to the part of the brain designed to be used with the left eye, and vice versa, right-hand images always appeared to come from the left and left-hand images always to come from the right. No amount of experience ever caused the toad to correct its mistake. It was obvious that the leftness and rightness of the vision were wired-in, and not learned, concepts.

Sperry's experiment was only one of many which have established that despite the ability of parts of the brain and nervous system to adapt and presumably modify themselves through learning, there is also a great deal of permanent wiring involved —many of the neuronal interconnections are formed during the embryological development period, and they are formed precisely, in the sense that the right neurons are tied to one another or to just the right sensory receptors or motor effectors. Thus the hope that learning alone would provide a solution for the interconnectivity problem has proved to be a false one. The problem is still with us.

A Hypothesis on Embryological Neuronal Wiring

No one yet knows for sure what physical or chemical factors control the details of embryological neuronal wiring. However, there is a general hypothesis that seems able to account for most of the observed facts. This hypothesis is suggested by the movements of cells in tissue cultures. Individual cells can be broken loose from the tissue of which they are a part and mixed with other cells in a liquid suspension. If in this way different types of cells—kidney and cartilage, say—are mixed together, a curious sorting out occurs. Under the microscope the cells, on encountering one another, are seen to slide over each other's surface in seemingly aimless fashion, but with the ultimate result that cells of the one type seek out one another and aggregate in one lump or layer, while cells of the other type form their own similar but separate association. Experiments have been performed with solutions containing several different types of cell, with similar results. Evidently, chemical substances in the cells result in attractive or cohesive forces specific to cell type that tend to cause like cells to stick together, unlike cells to remain unconnected.

The extension of this principle to the nervous system—the formation of the connections between retinal receptors and cortical neurons in the visual system, for example—involves the following line of speculation. The receptor neurons in the retina of the eye are assumed to contain two separate chemical ingredients that vary in concentration in accordance with the position of the neuron on the retina. One of these ingredients might appear in very small concentration in the rods and cones located at the extreme left-hand side of the retinal field, with the concentration of this ingredient increasing steadily across the retina to reach a maximum at the extreme right-hand side of the field. Similarly, the other ingredient might show concentration increasing progressively from the bottom to the top of the retina. With such an arrangement, the relative proportions of these two chemical ingredients in a given receptor neuron would provide an accurate indication of the position of the neuron on the retina, both left/right and up/down. Similar concentration gradients are presumed to exist in the interneurons of the optic system and the neurons in the brain with which the retinal receptors are ultimately connected. The embryonic growth process is assumed to be dynamic enough to cause each outgrowing nerve-cell fiber to wander close to a wide range of candidate terminating cells. By the operation of attraction or cohesive forces similar to, but much more specific than, those required to explain the sorting out of dissimilar cells in liquid suspension, the searching nerve fiber is assumed able to seek out and make connection with receiving neurons of similar concentrations of the two key chemical ingredients. In this way there results a continuous one-to-one correspondence between points on the retina and those on the visual cortex of the brain, and the picture we finally see is coherent and unscrambled.

Future work may or may not confirm the controlling role of connectivity ingredients in determining the built-in wiring of the nervous system. If not, however, it seems inevitable that some other effect will be discovered that produces the same result—an ability of growing nerve fibers, through physical or chemical interaction with the local environment, to search out and connect to other specific neurons or terminal organs. And the embryological prewiring in the nervous system will almost certainly turn out to employ mechanisms that differ only in degree, not in fundamental quality, from those which control the development of structure in the rest of the organism.

The Major Conclusion

It seems clear that the nucleic acid molecules exercise architectural control over a growing multicellular organism by a combination of direct and indirect methods. Their delineation of the structure of the enzymes, which in turn specify the chemical reactions permitted in the growing cells, is an important direct method of influencing the final outcome. However, we have seen that the physical and chemical conditions of the cellular environment are also important—that local physical forces distort and shape the growing tissue and that chemical agents in the surrounding fluids enter the cells to modify the enzymatic reactions and to trigger individual genes. To be sure, these local physical and chemical environmental factors are themselves results of the previous detailed development of the various parts of the organism, which in turn depended on the genetic mechanisms and the local environment, and so on back to the initial fertilized egg. Thus it is still correct to say that the genetic mechanisms exercise primary architectural control over all the design features of the organism. However, they do not do so by directly extracting from the nucleic acid book of instructions a completely detailed specification for each cell, which then controls a single-minded manufacturing operation to generate the specified product without regard for what may be happening in other nearby cells. Instead, the genetic mechanisms seem to have learned how to minimize their own detailed architectural chores by supplying to each cell a sort of "do-it-yourself kit" of chemical response features that enables the cell to develop properly through its automatic interaction with the local environment that it and its neighbors continually create and modify. The result is the scene of activity viewed by the embryologist—a delicately balanced and fantastically complex interplay between the standardized genetic mechanisms of the cells and their varied and changing surroundings. This is the secret of the diversity of structure and metabolism that makes multicellular plants and animals possible. Only by such methods is nature able to mold its raw materials into such an impressive end product as a living higher organism—a dandelion or a man.

A final comment may be in order about the evolutionary origin of the mechanisms on which the embryonic development of plants and animals is based. For despite the compelling nature of the evidence, normal human experience does not seem compatible with the conclusion that such remarkably complex and intricately interrelated mechanisms could have arisen solely through the blindly probabilistic workings of evolution. The antidote to such a feeling of skepticism is a reconsideration of the frequency of past occurrence of the typical evolutionary sequence: a small accidental change in structure of a nucleic acid molecule resulting in individuals with new anatomic or metabolic properties, their competition for survival, and the ultimate proliferation of the best-adapted species. The fantastically large number of such small refinements that must have taken place among quintillions of individual organisms during billions of years is also far beyond normal human experience. The essence of the theory of evolution is the balancing of the near inconceivability of its accomplishments against the correspondingly near inconceivability of its painstaking attention to detail.

BIBLIOGRAPHY

"Experimental Embryology," *Encyclopaedia Britannica*, ed. 14.

"Experimental Embryology," *Encyclopaedia Britannica*, 1962 ed.

Waddington, C. H., *New Patterns in Genetics and Development* (Columbia University Press, New York, 1962).

Wooldridge, D. E., *The Machinery of the Brain* (McGraw-Hill Book Company, New York, 1963), chap. 2, "The 'Schematic Diagram' of the Nervous System."

Wooldridge, D. E., *The Machinery of Life* (McGraw-Hill Book Company, New York, 1966), chap. 16, "The Structure of Plants and Animals."

part 2

Behavior

chapter 5

The Basic Behavior Mechanisms

To this point we have examined modern explanations for some of the physical aspects of life that once were thought to lie forever beyond the comprehension of the scientist. First, by studying the workings of the ordinary principles of physical science in organic material, we were able to understand the basic nature of the genetic mechanisms which exorcise the mystery from the lifelike *properties* of simple organisms. Then, in searching for an explanation of the *origin* of living things, we learned that the combining of ingredients and energy sources similar to those believed to have been a feature of the surface of the primordial earth automatically results in simple forms of the organic molecules on which these modern genetic mechanisms are based. We followed the inevitable organization of these materials into structures possessing primitive lifelike features. Finally, we considered the resulting competition of these structures for the same limited supply of raw materials. Thus we had our attention drawn to an important pattern in the operation of the physical laws, called evolution, which biases this kind of competition among different lifelike organizations of matter so as gradually to increase the relative population of those with structure and metabolism best contributing to survival and reproduction. Through the great power of evolution when operating for the several billions of years that geology allocates for the purpose, we found it possible to account for the structure and metabolism of today's plants and animals, with their vastly improved survivability and reproduction characteristics. In this way we were able to trace a continuous path of physical development all the way from the inorganic ingredients of the primordial atmosphere up to man himself.

To some, this is adequate proof of the validity of a powerful conclusion: *all* human characteristics must be completely accounted for by physical science. This would follow from the fundamental biological premise that the properties of an organism are determined by its chemistry and structure. However, despite the great advances in biology that have been made with the help of this premise, there has always been a reluctance to extend it to certain areas. Some of the most important characteristics of higher animals—behavior, intelligence, and consciousness—have been traditionally conceived by many to be of a nonphysical nature, not to be accounted for solely by the details of construction and metabolism of the organism. Therefore life scientists generally consider it important to develop independent evidence that the seemingly nonphysical attributes of organisms do indeed have a physical basis.

In this and the next chapter we shall examine the physical explanations for simple behavior. In Part 3 we shall extend our considerations to intelligent behavior and intelligence. Part 4 will deal with the even more difficult matter of consciousness.

The Nerve Cells

Since the plan of the next several chapters is to move gradually from simple to more complex considerations, we must commence by briefly examining the part of the anatomy that is known to be most directly involved in the control of behavior—the nervous system. Reference has already been made to the principal working element of the nervous system—the nerve cell, or *neuron*. In addition to recognizing that there are lots of neurons in a higher animal—ten billion in each human being—we shall need to know something about the properties of these tiny but important organs.

To begin with, neurons are essentially signaling devices. The most conspicuous structural feature of a neuron is a long, wirelike appendage called an *axon,* which carries signals from the main body of the cell to muscles, glands, or other neurons. Nerve cells come in a wide variety of sizes and shapes, but nowhere is more variety displayed than in the length of this axonal transmitting fiber. In the human being it ranges from a few thousandths of an inch up to a foot or more, depending on neuronal type.

The signal transmitted along the axon, befitting the wirelike appearance of the appendage, is electric in nature. Pulses of electricity, initiated in the cell body, travel along the axon to many tiny branching fibers near its far end. There the electric pulses trigger chemical reactions that discharge into the surrounding intercellular fluid certain special kinds of large molecules. If the surfaces of muscle cells lie close enough to a number of such discharge points to receive a substantial supply of these large molecules, the muscle will experience a resulting electrochemical reaction of its own which will cause it to contract and thereby perform some mechanical chore. In similar fashion a gland can be stimulated to display its characteristic chemical activity. And if the cell lying close to the terminations of the axonal fiber of the signaling neuron happens to be another neuron, then the chemical emissions from the first nerve cell may urge the second into a similar electrical activity, thereby propagating a signal along its axon to a still more remote part of the body. Neurons that, after activation by other neurons, cause muscular or glandular reactions are known as *effector* or *motor neurons;* those that are themselves activated by other neurons and, in addition, produce no other result than that of causing still other neurons to fire are known as *interneurons*.

But obviously not all neurons can be activated by other neurons. The process has to get started somehow. This is taken care of by specialized input devices known as *receptor cells*. These tiny organs take a number of forms, depending on the particular physical property they are designed to sense. In a touch receptor, for example, the displacement of an external hair compresses input appendages and thereby produces an electric potential which is converted in the axon into the usual train of electric pulses. Other receptors measure such quantities as pressure, warmth, cold, and displacement. In each case, evolutionary development has specialized the microscopic structure of the input end of the receptor neuron so that the standard electric signal is sent into the axon when changes occur in the particular physical property that the receptor is designed to detect.

In addition to receptor neurons measuring physical properties, the body contains many chemically activated receptors. The surface of the tongue and the membranes of the nose are lined with tiny nerve endings each of which is a full-fledged chemical analyzer that produces its standard electric output only when it comes in physical contact with some particular class of molecules.

Touch-sensitive and chemically sensitive receptors are sometimes found associated with other ingenious physical structures to form complex detecting systems. Consider, for example, the auditory mechanisms. There are no nerve cells that respond directly to sound waves. In order to hear, we must make use of a structure in the inner ear that analyzes incoming sound vibrations into a spectrum of mechanical displacements detectable by touch neurons. Specifically, the inner ear contains a long stretched membrane so designed that different spots on the membrane vibrate in response to different tones. Touch-sensitive neurons are arranged along this membrane so that the tiny hairs to which they are attached are distorted by the local vibration. The resulting pattern of electric pulses traveling along the axons of these touch receptors is what the brain interprets as speech, a symphony, or a baby's cry.

While we hear by means of touch, we see by means of chemicals. Reference has already been made to the rods and cones of the retina of the eye. These are receptor nerve cells that translate the pattern of light and shadow produced on the retina by the focusing action of the lens into the kind of signals required by the brain for implementation of the sense of vision. In view of the discussion to this point, it will surprise no one to learn that the signals generated in the eye are of the electric-pulse form that we have learned is standard in the nervous system. What may seem surprising, however, is that the rods and cones do not operate on a photoelectric principle in their conversion of light to electricity. Instead, they are essentially chemical detectors. They depend for their action upon substances in the retina that are decomposed by light, in much the same way that the silver compounds of a photographic plate are decomposed. It is the substances resulting from this decomposition, not the light itself, that produce the electrical effects in the neurons.

Much more could be said about the ingenuity of structure and versatility of operation that has been designed into nerve cells by millions of years of evolutionary refinement. In particular, many of the details of the electrochemical processes underlying their unusual properties are pretty well understood. For our purposes, however, the preceding general statement of neuronal characteristics should suffice. Let us now see if what we have learned is capable of throwing light on the behavior of higher animals.

Simple Reflex Action

The simplest of all behavior is *reflex action*. Therefore we might expect that our superficial understanding of neuron physiology should be most easily applicable to such automatic, unthinking responses. And this is true. Consider, for example, the familiar knee-jerk reflex, the involuntary upward kick occurring when a rubber mallet is applied briskly to the soft spot just under the kneecap of a crossed leg. A chain, or *reflex arc,* only two neurons long is involved in this process. The mechanical shock of the mallet generates electric signals in the receptor neurons that constitute the femoral nerve, and these signals are transmitted by the long axons of these receptor neurons to a region of the spinal cord. Here the electric signals are passed to effector nerve cells whose long axons proceed to the fibers of the quadricipital muscle of the leg, where the resulting release of chemical mediating substance contracts the muscle and produces the observed upward kick.

Another example: When a cold object touches the skin, a reflex arc through the spinal cord, this time involving a chain several neurons long, activates the muscles of the hair follicles to produce the familiar phenomenon of gooseflesh. At the same time a related neuronal circuit constricts the capillary blood vessels just under the skin at the site of the cold object, thus reducing the amount of cooling of the blood stream.

While the simplest reflex arcs, such as those described, employ the material of the spinal cord as the switchboard for interconnecting the receptor and effector neurons, those responsible for more complex control reactions generally involve some of the nerve cells in the brain. The *light-reflex arc* is of such a nature. This control circuit causes the pupils of our eyes to dilate in weak light and constrict in strong light. The neuronal arrangement involves a certain degree of sophistication, somewhat reminiscent of a modern electronic computer. For there is here no direct connection of receptor and effector neurons. Instead, before use is made of the electric indications of light intensity provided by the retinal neurons, their signals are merged or blended. This is accomplished by interneurons, each of which receives chemical stimulation simultaneously from the axonal terminations of many different receptor cells. In such an arrangement the strength of the signal that is propagated along the axon of the interneuron is a sort of average of the signals of the receptor neurons that provide it with input stimulation. The final result of several stages of such blending or integrating in the light-reflex circuit is an electric signal representative of the average light intensity over the entire field of view. This integrated signal appears at a point near the top of the *brainstem,* a deep-lying part of the brain which is in effect an upward continuation of the spinal cord. There connections are established with outgoing nerve fibers that proceed to the *constrictor pupillae* muscle of the eye. The size of the pupil is determined by the degree of contraction of this muscle, which in turn depends on the magnitude of the electric current carried by the fibers of its controlling nerve.

The way in which we are protected from overheating is another interesting example of a brain-controlled reflex. There are two principal mechanisms for dissipating the excess body heat resulting from warm weather or unusual exertion. One of these compensating mechanisms consists in evaporative cooling by perspiration; the other consists in enhanced thermal radiation from the blood, the circulation of which is increased by dilation of the network of subsurface blood vessels. The control center for these cooling mechanisms, like that for the light-reflex arc, is in the brainstem. However, the neuronal circuit for the temperature-control system is somewhat unusual, in that its sensing device is also located within the control center rather than in a peripheral position. There is a good reason for such an arrangement. The function of this control system is to maintain constant the temperature of the vital internal organs rather than, for example, that of the skin. But the most critical of all the temperatures of the body is that of the brain itself. Therefore, in this instance, the logic of evolution has resulted in locating the controlling sensors directly within the brain. Temperature-sensitive receptor neurons generate axonal currents that increase as the temperature of the blood in the brainstem departs from the desired value of 98.6°F. These electric signals, suitably blended, pass into the multiple nerve fibers that traverse the muscular walls of the blood vessels directly under the skin, as well as to the nerves that control the sweat glands. Thus when the temperature measurement indicates that the blood in the brain is getting warmer than it should be, the surface blood vessels dilate and the sweat glands perform their assigned functions. A temperature change of only a few hundredths of a degree in the blood coursing through the brain produces an observable response by the cooling mechanisms.

Coordinated Reflex Actions

In the interneuronal blending of input signals we have had our first exposure to the idea that the wiring of the nervous system does not exclusively consist of the direct interconnection of receptor and effector cells, as was once thought. Multiple interconnection of the neurons is also typical of reflexes involving coordinated responses to a single stimulus by a number of different muscles and glands. The startle reflex is a good case in point, for it is one we have all experienced. It can be produced by a sudden, unexpected loud noise. We close our eyes and duck our head, bend our knees, and bring our elbows in close to our sides. All the effector neurons initiating these muscular responses receive their stimulating inputs from chains of interneurons leading back to the same auditory receptors.

The patterns of coordinated muscular activity that are controlled by reflex circuits are sometimes quite complex. Breathing, for example, employs more than ninety muscles, which perform their contractions and expansions in suitable rhythm as a consequence of the electric impulses sent out from the brain to over a thousand individual effector neurons. Swallowing food is also an amazingly complex process. Muscles in the diaphragm and the tongue must perform a synchronized operation. At the critical moment, the soft palate must move back to protect the nose cavity, the cartilages of the larynx must displace themselves to close the windpipe, and the epiglottis must duck out of the way before the mouthful of food passes by. And just the performance by an uninterested reader of the simple act of raising his hand to his mouth and stifling a yawn requires the transmission from the brain of precisely related electric control signals to synchronize the contraction of fifty-eight different muscles working on thirty-two separate bones in the hand and arm, not to mention the thirty-one muscles of the face that move the features and produce its various expressions.

The Major Conclusion

There are several thousand reflex circuits in the human body. Automatic and unthinking, they keep the organism alive and healthy. The kind of behavior they represent, while occasionally complicated owing to the subtle effects produced by the cross-connection of large numbers of neurons, is clearly understandable in terms of the electrochemical properties of the interacting cells.

With this conclusion it might appear that we were ready to move on to more difficult subject matter in our search for evidence of the physical basis of behavior. But we shall first spend another chapter on automatic response—not in order to develop more detail about the physical processes involved, but rather in order to simplify our task when we come later to a consideration of intelligence. For much of what passes for intelligent behavior is revealed under close scrutiny to be the product of reflexes similar to those we have been considering. We need a more extensive understanding of the range of response that is possible through reflex action alone, in order not to waste our time later in complex explanations for behavior that is in fact automatic and unthinking.

The production of intelligent-appearing behavior by combinations of automatic reflex circuits is most easily studied in animals with simpler nervous systems than that of man. Therefore such organisms as sea urchins, insects, and birds will constitute the cast of characters for our next investigation. We shall defer to

Part 3 consideration of the role played by automatic unthinking reflex responses in causing human behavior to appear intelligent.

BIBLIOGRAPHY

Benzinger, T. H., "The Human Thermostat," *Scientific American,* January, 1961, pp. 134–147.

Brazier, M. A. B., *The Electrical Activity of the Nervous System* (ed. 2, The Macmillan Company, New York, 1960).

Fulton, J. F., *Muscular Contraction and the Reflex Control of Movement* (The Williams & Wilkins Company, Baltimore, 1926), pp. 34–37.

Galambos, R., *Nerves and Muscles* (Anchor Books, Doubleday & Company, Inc., Garden City, N.Y., 1962).

Wells, H. G., J. S. Huxley, and G. P. Wells, *The Science of Life* (Doubleday & Company, Inc., Garden City, N.Y., 1938), book 1, chap. 3, sec. 5, "Sensation and the Senses," pp. 111–127.

Wooldridge, D. E., *The Machinery of the Brain* (McGraw-Hill Book Company, New York, 1963), chap. 1, "The Electrical Properties of Nerves," and chap. 4, "Automatic Control Circuits in the Nervous System."

chapter 6

Automatic Behavior Patterns of Animals

Reflexes and Tropisms

Upon the approach of a possible enemy, a barnacle abruptly closes its shell, a tube worm snaps its exposed feeding tentacles back into its protective tunnel of sand, a sea squirt contracts into a gelatinous blob, burrowing bivalves withdraw their soft protruding siphons into the sand. A sea urchin turns its pointed needles in the direction of danger, and the pincerlike jaws at the base of the needles stand up, ready to seize any enemy that comes too close.

These have been found to be simple reflex actions. There are tiny photocells among the sensory neurons of the barnacle, tube worm, sea squirt, and bivalve, and the shadow cast by an approaching enemy causes these photoreceptors to generate their standardized trains of voltage impulses that stimulate the muscles employed in the resulting avoidance reaction. In the case of the more complex reaction of the sea urchin the receptor neurons are chemical rather than photoelectric: they "taste" the surrounding salt water for the characteristic "flavor" of an enemy. The completely automatic and local nature of the response is demonstrated by the fact that a tiny chip broken away from a living sea urchin's shell, with only a single spine or stalked beak attached to it, will show the same alarm and preparation.

In addition to reflexes, nature makes extensive use of *tropisms* in regulating the behavior of its simpler creatures. A tropism is an automatic response differing from other reflexes only in that it affects the movement of the complete organism. When an earthworm digs down and finds the moist decaying vegetation on which it thrives and at the same time avoids the surface where it might furnish a meal for a passing sparrow, this is not the intelligent, planned procedure that it appears to be. The muscles that turn the front end of the worm and thereby determine its direction of locomotion receive electric control signals from photosensitive receptors on either side of the head. As a result, the earthworm automatically turns away from the light and heads in the direction that equalizes the amount of illumination received by the left- and right-hand photoreceptors. This causes the worm to travel toward the darker regions, where it finds food and safety. The completely machinelike, unreasoning nature of this performance has been nicely demonstrated by exposing a worm simultaneously to two separate sources of light of controllable intensity and observing that the path followed is always one that orients the worm, in accordance with the positions and relative intensities of the lights, to equalize the amount of illumination of its two photoreceptors; and this occurs even though it may impel the animal along a course opposite to the one in which the proper conditions for food and safety are to be found.

The machinelike nature of the reflexes and tropisms that so extensively regulate the behavior of the lower animals was not appreciated as soon as it might have been by workers in the field. This was probably because these responses do not have the precision and detailed reproducibility that is observed in the commoner reflexes of higher animals. An earthworm, when exposed to light, does not instantly snap into an opposite heading and pursue a straight course steadily away from the source of illumination. Instead, its trajectory is modulated by wormlike twistings and turnings; these deviations may become particularly extensive if, for example, it is necessary for the worm to avoid an obstacle that lies in its path. The reason for such imprecise response is that most tropisms do not provide such an overriding source of control voltage to the organism's effector mechanisms as to overshadow completely the effects of other sources of command signals. In addition to the neuronal connections causing the muscles to tend to turn the worm away from a source of illumination, there is also a built-in control pattern that provides for detouring around obstacles. The actual muscular response at any given instant is determined by the combined effects of these and several other reinforcing or competing built-in reflex or tropism mechanisms.

Another kind of complicating factor that impeded early understanding of reflex behavior is demonstrated by the sea urchin. Its spines usually display a certain random motion even when no stimulus is present, and their orientation toward an approaching enemy may also include some continuing restlessness. The disturbing control signal involved here is closely analogous to the noise on the circuits of the communications engineer. For the nerve cells of these lower animals, like many of those in the human body, do not always wait to receive a specific stimulus before "closing the switch" and sending voltage pulses out over the axon. Instead, there is a certain amount of random firing of the neurons. In the complex nervous systems of the larger animals, so many neurons must act cooperatively to produce significant movement of principal organs that the random firing of a few neurons cannot produce a conspicuous result (although an occasional flicker of an eyelid or twitch of a muscle may be due to this cause). In the small and primitive animals that we are now dealing with, however, there may be only a few interconnected neurons in the circuit that controls a major element of the body. Under such circumstances, the random firing of one or two neurons can produce observable restless movement of the affected part.

Complex Tropisms

Tropisms that are themselves modified by environmental factors can increase immeasurably the apparent purposefulness of behavior of a simple organism. For example, the larvae of barnacles possess a temperature-reversible tropism that causes them to seek light when cold and avoid it when warm. This simple mechanism contributes to their behavior an appearance of free will, as they "decide" whether they want to swim toward or away from the surface of the sea. Similarly, many aquatic crustaceans, such as the water flea *Daphnia,* tend to swim downward in a bright light and upward in darkness.

An interesting tropistic mechanism causes the caterpillars of the goldtail moth (*Porthesia chrysorrhoea*) to leave their hibernating nests in early spring and crawl to the only parts of the shrubs where their leafy food is to be found at that time of the year. The tropism involved is one whereby an adequate amount of warmth automatically causes the caterpillar to leave its nest and crawl toward the light; an experimenter can induce it at any time simply by applying heat. This tropism results in the caterpillar climbing as high as it can go, which is to the top of the shrub where the new growth of green leaves first emerges early in the spring. However, if other effects than this simple tropism were not operating, the caterpillar would be in difficulty as soon as it had eaten the green leaves at the top of the shrub, for its food from then on would have to be found at lower levels; reaching such levels would be in conflict with a tropism continuously impelling it upward. But the upward-climbing tropism operates only when the caterpillar is hungry. Therefore, having eaten, it is free to creep in any direction and will eventually wander down and find the new leaves as they begin to open.

As with all tropisms, the behavior of the goldtail moth is completely unreasoning. For example, if caterpillars are taken as they are leaving the nest and put into a glass tube lying near a window, they will all collect in the end of the tube nearest the light and stay there. If a few young leaves from their food shrub are put at the other end of the tube, farthest from the light, the hungry, unfed caterpillars will remain held captive near the lighted end of the tube, and there they will stay until they starve.

Tropisms, like the reflexes of higher animals, are direct consequences of the way the nerves and muscles are put together. Usually they contribute to the health and well-being of the organism. This, of course, would be an inevitable consequence of evolutionary selection; creatures with tropisms lessening their chances of survival would presumably not have won out in the struggle for species existence. It is only when the animal is placed in historically unusual circumstances that tropisms can work against survival, as in the example of the caterpillars in the lighted tube. Similarly, when moths and other phototropic insects are irresistibly drawn to destruction in a flame, this is only a result of a normally unimportant feature of their construction that superimposes upon the random motions characteristic of their aimless flight a steady pull toward the light. And the prawns that accumulate around the positive pole of a pair of electrodes placed in their tank do so because of another anatomic accident that weakens the effectiveness of their muscles when electricity passes through them in one direction and strengthens the effectiveness when the current is in the other direction. If open flames in the forests and electric currents in the seas had been important features of nature in the past, it is likely that moths and prawns would not have survived the evolutionary processes.

Although combinations of tropisms and simple reflexes seem to account for a surprisingly large fraction of the behavioral responses of lower animals, such simple nerve circuits do not by any means constitute the extent of nature's provision for unthinking, unlearned, but constructive behavior. Let us now pass to the evidence for the existence of much more complex patterns of neuronal interconnection, whereby elaborate sequences of interrelated actions may be called forth by the receipt of suitable stimuli.

Stored Programs of Behavior

Everyone has noticed that different kinds of birds move across the ground in different ways. One species may progress by hopping on both feet, another by walking, one step at a time. While the two species of birds may be physically quite different, this difference in their method of locomotion is not a consequence of muscular requirements or any other special aspects of their construction that we ordinarily consider to be physical. The bird that hops could just as well have been designed to walk, and the bird that walks could just as practicably have been designed to hop. What is involved is a difference in behavior. Birds of the one species inevitably become hoppers, and birds of the other species inevitably become walkers. They have no capability of changing their behavior patterns; the hoppers could no more walk than they could change their size and coloring, and the walkers will go through life taking their steps one at a time.

As another example of inherited behavior, consider the nest-building habits of the four groups of birds that constitute the conventionally recognized family of titmice (Paridae). One group always nests in hollow trees or other cavities; a second group builds an oval nest with lateral entrance in bushes and trees; a third group builds a peculiar retort-shaped nest of plant down worked into feltlike consistency; the fourth group builds a stick nest with a lateral entrance. While these birds are apparently physically identical, their innate, unlearned, nest-building habits serve to identify them as different species as clearly as though the four groups were marked with red, yellow, blue, and green feathers.

Species-connected peculiarities of behavior also appear in the language of animals. Thus while there is considerable similarity among the cries of alarm of all gulls, the number, pitch, and frequency of the staccato cries that constitute such calls vary among species.

Indeed, it is not unusual for behavioral characteristics to provide important clues to the proper classification of species. For a long time, the group of desert birds called sandgrouse (Pteroclididae), which has downy young greatly resembling young grouse (Tetraonidae), was considered to be a member of a closely related family. Later, more careful analysis of physical characteristics led to the suspicion that sandgrouse were more closely allied with pigeons. This suspicion was finally confirmed and the classification corrected by employment of a behavior characteristic. While nearly all birds scoop water up with their bills and then let it run down into their stomachs by lifting head and neck, pigeons have a very different drinking behavior: they stick their bills into the water and simply pump it up through the esophagus. The fact that sandgrouse are the only other birds with this behavior strongly reinforced the anatomic findings which placed them next to pigeons.

Inherited behavior patterns are by no means confined to birds. Separate family classification has been assigned to different groups of grasshoppers largely because of differences in methods of cleaning the antennae. Thus the Acrididae clean their antennae by pulling them between the leg and the ground. Physically similar, the Tetrigidae family clean their antennae by stroking them with the legs, which in turn are cleaned by being pulled through the mouth. Similarly, in the sea some species of hermit crabs have the instinct to find cast-off shells as houses for their unprotected abdomens, while other species protect themselves by holding stinging sea anemones in their claws.

At one time, it was assumed that indoctrination of the young by their parents was responsible for all such species-specific behavior, since the ability to learn is possessed by so many creatures. However, attribution of the examples considered to be heredity rather than learning is consistent with current knowledge. Indeed, careful observation has shown that many relatively complex habits must exist in animals at birth as finished and complete patterns of behavior. For example, immediately after hatching certain birds will automatically crouch down in the nest when a hawk passes overhead. This is not simply a response to a dark object in the sky. The shape must be hawklike; a robin can pass overhead without evoking the slightest reaction. Consider also the so-called thermometer bird, or bush turkey, of the Solomon Islands. It lays its eggs in a heap of mixed plant material and sand, with all the eggs arranged to lie with the blunt end upward. Each chick, on breaking out of the blunt top of its egg, wriggles and struggles in such a way that its stiff feathers, which point backward, gradually cause it to work its way up to the top of the heap. On reaching the surface, the chick dashes cross-country to the shade of the nearest undergrowth. Certainly no learning is involved in this response pattern of the newly hatched chick. Similarly, a female canary that has been isolated from birth builds a nest competently the first time suitable material is presented and the occasion arises. And a caterpillar, when it is about to pupate, spins a cocoon. It may never have seen another caterpillar or cocoon, and yet it automatically sets about to construct an edifice that, when analyzed, is seen to be a masterpiece of engineering.

Although such observations show conclusively that learning from experience is not occurring, the inheritance at birth of detailed and purposeful behavior patterns seems so different from anything we human beings experience that it is necessary for us to continue to fight against the tendency to imagine that reasoning intelligence is involved. We must therefore not ignore the evidence on this point. Consider again the thermometer bird, which on emerging from the egg executes exactly the kind of wriggling motion needed to bring it to the surface of the heap and then changes to a new mode of motion to bring it to the protection of shade. If the chick, after having emerged, is once more dug into the heap, it is quite incapable of coming out again and stays there struggling ineffectively until it dies. Its movements are now of the type suited for running to shade, not of the type that will bring it to the surface. And the caterpillar that builds such a wonderful cocoon displays the completely automatic nature of its performance if it is interrupted in the middle of its task and the half-finished cocoon is removed. It does not start again from the beginning, but spins only what remained for it to do, in spite of the fact that the resulting half cocoon is completely useless for protection. Similarly, the octopus that so "intelligently" builds a stone wall behind which it can hide unseen will with equal vigor construct the wall out of transparent pieces of glass if this is the material that happens to be handy.

In the light of present knowledge, we can only conclude that these specific behavior patterns result from correspondingly specific neuronal wiring diagrams that are built into the organisms at birth. The same kind of kinetic embryonic forces that determine all the millions of physical details of an animal also determine the detailed patterns of interconnection of the neurons in its brain; and the patterns of behavior that result are as unique to the species as obviously physical characteristics like size, shape, and color.

"Intelligence" of Insects: Triggering of Stored Subroutines

A special challenge to this hypothesis is provided by the elaborately organized activities of some insects, such as ants, termites, bees, and wasps. For years, man has been fascinated by these insects. He has read into their organized behavior strong elements of similarity to the reasoning processes of humankind. Let us see if this interpretation survives close analysis, or if our concept of permanently wired-in neuronal circuits again appears to fit the facts.

Consider, for example, the solitary wasps. When the time comes for egg laying, the wasp *Sphex* builds a burrow for the purpose and seeks out a cricket which she stings in such a way as to paralyze but not kill it. She drags the cricket into the burrow, lays her eggs alongside, closes the burrow, then flies away, never to return. In due course, the eggs hatch and the wasp grubs feed off the paralyzed cricket, which has not decayed, having been kept in the wasp equivalent of a deepfreeze. To the human mind, such an elaborately organized and seemingly purposeful routine conveys a convincing flavor of logic and thoughtfulness—until more details are examined. For example, the wasp's routine is to bring the paralyzed cricket to the burrow, leave it on the threshold, go inside to see that all is well, emerge, and then drag the cricket in. If the cricket is moved a few inches away while the wasp is inside making her preliminary inspection, the wasp, on emerging from the burrow, will bring the cricket back to the threshold, but not inside, and will then repeat the preparatory procedure of entering the burrow to see that everything is all right. If again the cricket is removed a few inches while the wasp is inside, once again she will move the cricket up to the threshold and reenter the burrow for a final check. The wasp never thinks of pulling the cricket straight in. On one occasion this procedure was repeated forty times, always with the same result.

In connection with the interneuronal averaging of incoming electric signals we have already had occasion to remark a certain resemblance between the properties of complex nerve circuits and those of electronic computers. In Chapter 8 this theme will be expanded by the discussion of an important set of basic construction and performance characteristics that are common to nervous systems and computers. At this point, therefore, it is pertinent to call attention to the sense of familiarity felt by the computer scientist on confronting the type of behavior that we have just been considering, for it has the earmarks of a set of subroutines recorded in the permanent memory system of a computer and called into play by the appearance of certain conditions of the input data. In the instance of the solitary wasp, it would appear that some triggering mechanism, perhaps the physiological state of the female, sets into motion the series of subroutines associated with the preparing of a nest and the laying of eggs. The first subroutine called forth is the preparation of a burrow. The completion of this subroutine is the trigger for the next, which consists in the searching down of a particular species of cricket and paralyzing it. This in turn is the trigger for the next act in the sequence—bringing the cricket to the threshold of the burrow. The presence of the cricket at the threshold of the burrow is the signal for the wasp to go inside for a last check around. Emergence from the burrow and finding the paralyzed cricket still there is the signal for pulling the cricket into the burrow, and so on. Just as in the design of complex programs for electronic digital computers, subroutines appear to be stored and triggered into operation by the particular combinations of stimuli called for by the stored control program of the mechanism.

This concept of stored subroutines that are triggered by specific stimuli goes a long way toward accounting for the surprising variety of detailed inherited behavior patterns exhibited by insects. For example, a bee that has found food will, on its return to its hive, execute a characteristic wagging dance by means of which the direction, distance, amount, and quality of the food source are communicated to the other bees. But the worker will perform its dance as artistically in the absence of other bees as in the presence of an audience. All that is necessary to trigger the performance is stimulation of its antennae.

And the social insects—ants, termites, and bees—all appear to be patriotic, for they will drive out and frequently sting to death individuals from other hives. But the trigger is odor. All is changed if the interloper is protected long enough to acquire the scent of the new hive. In fact, suitably odor-conditioned insects of entirely different species will frequently be allowed to live indefinitely in a colony of ants or termites.

Indeed, the senses of odor and taste are involved in many insect trigger mechanisms. The great devotion to their queen of termite workers, hundreds of whom are generally in attendance upon her, appears to be a simple consequence of the fact that she exudes an especially rich and fatty secretion; their apparent attentions consist in licking her to get something for themselves, sometimes so violently that they rasp holes in the royal side. Superficially similar, but for an entirely different purpose, is a detail in the mating activities of certain species of spiders, whereby the male is stimulated to suitable attentions toward the female by an attractive substance that she exudes over her body.

Because of the specificity of the trigger mechanisms, these innate response patterns are, by human standards, ridiculously rigid and inflexible. Thus a male nocturnal moth may fly unerringly to his mate for a distance of more than a mile; yet if the feathery antennae that serve the male as sense organs are cut off, he is not only incapable of finding the female but also incapable of mating if placed alongside her. The trigger for this act apparently is the smell stimulus normally supplied to his sensory antennae. The odor that so stimulates the male moth is generated by two little scent organs located near the tip of the female's abdomen. These organs can be cut out without particularly inconveniencing the female. If they and the operated female are then put in a cage with a normal male not deprived of his antennae, his built-in pattern of mating actions will be triggered but will be entirely directed toward the source of the stimulus; he will make vain attempts to mate with the two little scent glands but will entirely ignore the female.

An invertebrate animal may starve to death in the midst of plenty if the particular plant or animal material that serves as food for its species happens to be missing. And an insect may doom its race (or at least its local colony) to extinction because of the absence of the particular stimulus that is required to set off its pattern of nest building or egg laying. While a *Sphex* wasp provides its grubs with crickets, the *Ammophila* must find and paralyze a caterpillar before it can continue with its nest-building and egg-depositing routine. The *Sceliphron* wasp recognizes only spiders as suitable larva food, and the *Podium* uses roaches. The *Pronuba* moth can lay its eggs only in a yucca plant, and a species of Trinidad mosquito can be triggered to deposit its eggs only by the presence of the leaves of a bromeliad plant floating in a pool of stagnant water.

So there is, indeed, abundant evidence of the existence in insects and lower animals of premanently wired-in inherited patterns of behavior. Ranging from simple reflexes and tropisms up through complex patterns of multistage behavior triggered by specific stimuli, these automatic, machinelike responses have many of the earmarks of the library of subroutines that can be stored in the memory of an electronic computer and triggered into operation by the occurrence of a prescribed series of relationships among the data supplied to the computer by its input devices.

Genetic Control of Behavior: Evolutionary Hypothesis

Before leaving the subject of automatic behavior patterns, we should assure ourselves that their origin, like their detailed properties, is reasonably explained by the normal operation of the ordinary laws of physical science. In view of what has gone before, the pertinent argument can be simply stated. It starts again with the observation that these behavioral characteristics are consequences of the physical construction of the organism—in this case, of the specific way in which the neurons are interconnected. The specifications for these interconnections, like those for the more obviously physical characteristics of the organism, are carried by the genes. The same built-in mechanism by means of which these nucleic acid genetic blueprints, in their interaction with the surrounding chemical materials, compel precise adherence to their design specifications in the embryonic growth of the muscles or visceral organs automatically and accurately wires the neurons into the precise control circuits required by heredity. Thus it must follow that the same evolutionary processes that determine physical structure also control the development of inherited behavior characteristics: a Change in the behavior pattern of an individual is occasionally caused by a mutation in a gene that helps specify the wiring of the neurons. If the change in behavior results in increased survivability, the descendants of the affected individual will grow more numerous from generation to generation, and the new behavioral characteristic will ultimately become typical of the species.

At least in the case of simple reflexes and tropisms, it is as easy to visualize the workings of such evolutionary principles upon behavior as upon the more obviously physical characteristics of the species. A mutation in the neuronal scheme of the nocturnal moth resulting in a more effective interconnection between the sensory neurons responsive to the scent of the female and the effector neurons controlling the flight muscles could easily result in a new strain of moths of enhanced mating effectiveness. If individuals of this new strain produced on the average 20 percent more progeny than the nonmutated individuals, only four generations would be needed for a colony of moths of the new strain to attain twice the population of a colony of old-style moths of initially the same size. Of course, in an actual situation in which the new strain starts out as a tiny minority element in an unmodified population, with no artificial forces at work to speed the process by controlled interbreeding of the individuals having the desired characteristics, much longer periods of time are required by the evolutionary processes to produce noticeable changes. On the other hand, we must remember that the population in any one generation of insects and some of the other lower animals may run into many billions and also that the life span of an individual in these lower forms is so short that a complete generation may be compressed into a few weeks or months. Therefore it is easy to imagine that evolution, in the millions of years it has had for working its developmental wonders, should have found its random statistical methods adequate for refinement to today's state of the reflexes and tropisms that underlie so much of the behavior of the simpler forms of life.

Genetic Control of Behavior: Direct Evidence

Evolutionary hypotheses are difficult to confirm by direct observation. However, no such uncertainty need extend to the interpretation of the experimental evidence for the genetic control of behavioral characteristics in existing organisms. For example, strains of rats have been isolated that exhibit marked differences in their behavior toward strangers. In one genetic strain, extreme aggressiveness and hostility are displayed; in another strain of rats that in other respects appear identical, the behavior is docile. These behavioral characteristics breed true, just as do the physical characteristics determined by the genes. In another experiment, a particular gene of a fruit fly has been identified that controls one of the details of behavior during the act of mating. Specifically, a modification in this gene decreases the strength and duration of the vibrations of the wings and antennae with which the male fly caresses the female at a certain stage in the courtship procedure.

Evidence for the hereditary control of the properties of the neurological circuits is also supplied by human victims of genetic defects. Phenylketonuria, a defect that often causes mental retardation, has been determined to be hereditary in nature. So is juvenile amaurotic idiocy. There are also two different types of muscular tremor, each highly specific to local population groups in New Guinea, a progressive dementia found in certain islands of the western Pacific, and so on.

With the growing understanding by physicians of the possible genetic significance of the symptoms of their patients, clues to the existence of hitherto unrecognized hereditary diseases are being discovered at a rapid rate. Because so many more human beings than animals come under medical scrutiny, it is from such sources, rather than from animal experiments, that most new evidence is coming for neurological genetic defects.

The Major Conclusion

We have learned that the kind of behavior that is completely determined by fixed interconnections among the neurons is not always as automatic in appearance as we might have expected. When a single organism possesses a number of reflexes, tropisms, and stored subroutines, each triggered by its own prescribed input stimuli, many of which can occur concurrently and sometimes in competing fashion, the complexity of behavior of the animal is much increased, and the machinelike nature of its responses is obscured. The superposition on these responses of a certain degree of randomness, arising from the spontaneous firing of the neurons, also enhances the lifelike character of the behavior. Thus the specific built-in wiring diagram of the neurons is capable of accounting for a large part of the behavior of lower animals.

And the origin of the behavior-determining pattern of neuronal interconnection, like the origin of all other structural characteristics, appears to be accounted for satisfactorily by the inextricably intertwined concepts of evolution and genetics. The same evolutionary/genetic mechanisms, of course, must also be responsible for the development of the many reflex mechanisms in the higher animals which contribute so much to survival capability and which we have seen are so similar in nature to the inherited behavior patterns of lower forms.

We are now ready to move on to a more difficult topic—the explanation of intelligence by means of the ordinary physical laws of nature.

BIBLIOGRAPHY

Bastock, M., "A Gene Mutation Which Changes a Behavior Pattern," *Evolution,* vol. 10 (1956), pp. 421–439.

Dobzhansky, T., "Eugenics in New Guinea," *Science,* vol. 132 (1960), p. 77.

Eibel-Eibesfeldt, I., "The Interactions of Unlearned Behaviour Patterns and Learning in Mammals," in *Brain Mechanisms and Learning,* ed. by Fessard, Gerard, Konorski, and Delafresnaye (Charles C Thomas, Publisher, Springfield, Ill., 1961), pp. 53–74.

Emerson, A. E., "The Evolution of Behavior among Social Insects," in *Behavior and Evolution,* ed. by Roe and Simpson (Yale University Press, New Haven, Conn., 1958), pp. 311–335.

Mayr, E., "Behavior and Systematics," in *Behavior and Evolution,* ed. by Roe and Simpson (Yale University Press, New Haven, Conn., 1958), pp. 341–362.

Pittendrigh, C. S., "Adaptation, Natural Selection, and Behavior," in *Behavior and Evolution,* ed. by Roe and Simpson (Yale University Press, New Haven, Conn., 1958), pp. 390–416.

Thompson, W. R., "Social Behavior," in *Behavior and Evolution,* ed. by Roe and Simpson (Yale University Press, New Haven, Conn., 1958), pp. 291–310.

Thorpe, W. H., "Some Characteristics of the Early Learning Period in Birds," in *Brain Mechanisms and Learning,* ed. by Fessard, Gerard, Konorski, and Delafresnaye (Charles C Thomas, Publisher, Springfield, Ill., 1961), pp. 75–94.

Wells, H. G., J. S. Huxley, and G. P. Wells, *The Science of Life* (Doubleday & Company, Inc., Garden City, N.Y., 1938), book 8, chap. 1, "Rudiments of Behavior"; chap. 2, "How Insects and Other Invertebrates Behave"; and chap. 3, "The Evolution of Behavior in Vertebrates."

Wooldridge, D. E., *The Machinery of the Brain* (McGraw-Hill Book Company, New York, 1963), chap. 5, "Permanently Wired-in Behavior Patterns of Lower Animals."

part 3

Intelligence

chapter 7

Two Working Hypotheses

Part 2 has demonstrated that much animal behavior, despite its intelligent appearance, is the result of automatic reflexes arising out of permanent interconnections among the cells of the nervous system. The reader has not erred if he has concluded that automatic mechanisms might also play an important role in some of the human behavior we call intelligent. Indeed, in this and the next four chapters we shall examine evidence that in general, the processes underlying intelligence differ only in degree, not in essential quality, from the automatic processes underlying simple behavior.

But human intelligence involves thought, and thought, unlike the biological-research results that have thus far been featured in this book, is a subject on which each of us feels himself to be an expert. Every adult mind is a laboratory in which thousands of hours have been spent in the practice and observation of intelligent thinking processes. And there is a quality to this mental experience that *feels* irreconcilable with the notion that thought is the orderly and predictable consequence of the workings of the ordinary laws of physical science among the axons and cell bodies of the neurons. So general is this feeling that it would probably be futile to proceed with our physically based considerations without first dealing with it. Thus the purpose of this chapter is to examine some arguments suggesting that our subjective feelings may be misleading and that the idea of a physical explanation of intelligence and thought may not be as irreconcilable with our own mental experience as we think it is. If these arguments are effective, we shall then be able to go on to an examination of the scientific evidence for the physical basis of intelligence, with confidence that it will receive objective consideration by the reader.

Our intuitive disbelief in the purely physical origin of thought and intelligence seems to be based on two types of recurring subjective observation. These will be successively discussed in the next two sections.

The Apparent Freedom of Thought

Probably we would all be more receptive to the idea that thought is exclusively the result of the operation of physical law in the neuronal material of the brain if we were more aware of structural detail in our mental activities. In particular, if in thinking we were always conscious of a completely continuous, step-by-step progression of images and ideas, each followed by another only slightly different from the preceding, we would find it easier to believe that the underlying processes are of the detailed cause-and-effect nature characteristic of physical mechanisms. But this is not how thinking seems to take place. Instead, it appears to possess an independence of its own. Thoughts "come to us" in a remarkably unheralded manner. Although a suddenly appearing memory or idea usually has some features of resemblance to the idea that immediately preceded it, there is nothing like detailed continuity in our stream of thought; sometimes it is almost as though we were present in the audience while the poorly connected scenes of an inexpertly written play were caused by an invisible director to unfold before our eyes.

Even in the disciplined activity of problem solving, our mental processes appear to be only partly controlled by cause-and-effect relationships, for successive steps in the solution of the problem at hand rarely come to us in closely sequential logical order. Instead, our thought processes always seem to have enough of a "mind of their own" to force digressions to consider offshoots of the main problem, if not to deal with entirely unrelated subjects.

Such spontaneity certainly seems incompatible with the idea that thought is the result of the step-by-step, cause-and-effect processes that would have to characterize the electrochemical workings of the neurons. However, there is a way of reconciling these observations with the continuous and predictable performance we expect of physical mechanisms. The applicable argument can be developed by further use of the analogy of a playgoer in a theater.

This time our imaginary play will be a good one—good at least in the sense that each episode follows with perfect logic and continuity from what has gone before. To make the point as strongly as possible, we may go so far as to postulate that the play has been plotted by an electronic computer, programmed so as not to introduce the slightest spontaneity or discontinuity into scenery or action. This might result in our "good" play actually being remarkably dull, but it will still be a good play for our purpose. The point is that, despite the ultimate in logical continuity of its structure, the play could yet have the appearance of spontaneity to certain playgoers. A very sleepy member of the audience, for instance, might form such an impression from the bits of scenery and action on display during his occasional periods of consciousness. And if, instead of merely sleepy, the playgoer were a certain type of epileptic, subjcet to frequent periods of unconsciousness without knowing it, the result would surely be a completely honest belief that the play itself consisted of spontaneous and disconnected episodes.

The possible analogy with our thought processes is obvious. Could it not be that thought seems free and spontaneous only because most of the activity of thinking is unconscious and that only occasionally does a part of the action flash across our screen of awareness? Such a hypothesis is of course more directly suggested by the analogy of the epileptic than by that of the sleeper. For one of its consequences would be that we too would remain unaware of the incompleteness of our attention and would therefore attribute the discontinuity observed to the action watched rather than to the intermittancy of the surveillance.

Of course, the mere formulation of a hypothesis does not prove its correctness. But the evidence that may be capable of establishing a positive case for physical theories is to be set forth in the next chapters. Our present need is only to set aside subjective blocks to the objective consideration of such evidence. All that is required at this point is to establish that our mental experience does not *necessarily* preclude a physical explanation of thought. And the above hypothesis does this. If we accept as possible the idea that we are aware of only a part of the activities underlying thinking, then we have no reason to deny the further possibility that a complete description of those activities would show an absolutely continuous, cause-and-effect chain of underlying events. Therefore, in the following discussion, the strong feeling we all have that thought is free and independent will not be allowed to limit the scope of the investigation.

The Nonphysical Feel of Consciousness

We come now to an even greater subjective impediment to progress toward a physical explanation of intelligence—the existence of consciousness itself. Surely this personal sense of awareness that illumines some of our intellectual activities is qualitatively unique, not to be derived from the impersonal quantities, such as gravity, mass, and electricity, that form the subject matter of physics and chemistry. Since consciousness is involved in thinking, is it not therefore obvious that there can be no purely physical explanation of intelligence?

This objection is a sound one. In a strict sense it is indeed impossible to treat the subjective aspects of conscious intelligence in purely physical terms. At least it is impossible in terms of what are today the generally accepted laws of physics. This fundamental problem will be attacked in Part 4, where a solution based on the expansion of the subject matter of science to include conscious phenomena will be examined. Meanwhile, there is a working hypothesis about consciousness that should sustain us until we are able to consider it more fully. The hypothesis is that it is a passive phenomenon, that it doesn't *do* anything. This is by no means equivalent to denying the existence of consciousness. Instead, it amounts to limiting its role to that of a sort of window through which we can observe a part of the workings of the brain without interfering with the orderly operation of the machinery we are watching.*

Again, although the proposed hypothesis is known to be consistent with the developments to follow, no proof of its validity is offered at this point. The reader is asked only to agree that the hypothesis *could* be correct. We can then go on and objectively consider the evidence that intelligent activities can be completely explained in terms of the physical properties of the neuronal mechanisms. If this evidence is convincing, it will provide support for the hypothesis assigning a passive role to consciousness. If in addition we later learn there is a means of integrating a passive consciousness into a logically expanded concept of science, then we shall feel even more confidence in the validity of the hypothesis.

In summary, in return for the promise that a means of relating consciousness to basic science will be set forth later in the discussion, the reader is asked to give tentative acceptance to two work-

---

* The idea of the passivity of consciousness is of course an old one. What is not old is the strength of the modern evidence supporting it, which we are about to consider.

ing hypotheses: (1) we are conscious of only a fraction of our thought processes; and (2) consciousness is completely passive. If such assumptions are admitted to be possible, there is no a priori reason why a physical explanation of intelligence may not be found. The evidence supporting such an explanation constitutes the subject matter of the next four chapters.

chapter 8

Computers and the Brain

During each second of the seventy-year life span of the average human being, tens of millions of pulses of electricity arrive over long fibers from millions of sensory neurons to bring his brain information about the internal conditions of his body and its external environment. Special receptors in muscles and glands regularly report the configuration and state of activity of his parts and organs; optic nerve cells send in information about the surrounding patterns of light; neurons in the auditory system provide data on the ambient sounds; olfactory, taste, touch, temperature, and pressure-sensitive neurons continually transmit electric signals related to other aspects of his environment. And during each same second, tens of millions of pulses of electricity also leave his brain over the long fibers of millions of effector neurons to control the glands and muscles that in turn regulate his internal health and external activity. If he is Paderewski, the external activity resulting from the complex pattern of outgoing nerve signals may include the extraordinarily skillful manipulation of the keys of a piano. If he is an Einstein, the conspicuous output may be the muscular control of a pen to write, or of the throat muscles to speak, profound new scientific deductions. Explanation of such phenomena, in our terms, requires no less than the demonstration that *the lifelong sequence of such external, intelligent actions is completely and in detail the consequence of the operation of the ordinary laws of physical science in the material of the brain cells, under the continuing stimulus provided by the lifelong sequence of incoming neuronal signals.*

One of the striking curiosities of modern biology is the fact that nearly all of the significant evidence for such a physical basis of intelligence has emerged from research in a nonbiological field: computer science. To make the results of this research logically available to our present considerations, we must first examine the case for the family relationship of electronic digital computers and brains.

The Computer/Brain Hypothesis

Nothing is easier, or has more frequently been done, than to draw a certain kind of parallel between digital-computer installations and the nervous systems of men and animals. Consider, for example, an industrial process-control computer system—one which directs the activities of, say, a chemical plant. There is an obvious analogy between the devices that derive electric indications of pressure and temperature for use by the central computer of such a chemical control installation and the pressure and temperature receptors in the periphery of the body, with their nerve-fiber connections to the brain. And a similar analogy exists between the actuators that open and close valves to implement the commands of the computer and the muscles by means of which the body obeys the dictates of the brain. Inevitably, therefore, we find it easy to infer a general analogy between the centralized process-control computer that receives data and generates commands and the centralized brain, which receives similar-seeming data and generates similar-seeming commands.

This kind of analogy provides encouragement for a scrutiny of the two kinds of central data processors in search of significant elements of similarity. And encouragement is needed, for there is much about the electronic digital computer which seems irrelevant to biological structures. Its punched cards, rotating disks, and flashing lights appear as far removed as anything could be from the quiet assemblage of interconnected nerve cells that we believe to constitute the working machinery of the brain. But modern engineering has provided many examples of devices that, while markedly different in physical structure, are functionally equivalent. The same symphony concert can be recorded either on a magnetic tape or on a grooved wax disk, and both may ultimately be supplanted by an entirely different recording medium as technology develops in directions we cannot now foresee. In general, the particular devices and techniques our engineers find most suitable for attaining a functional result are transient and changeable. The important similarities between computers and brains, if there are any, must relate to their basic principles of operation rather than to the kinds of components used in their construction.

It is therefore pertinent to ask, "What is the *essence* of an electronic digital computer? What is the fundamental nature of its operation?" Fortunately, there is a clear-cut answer. When its performance is reduced to basic essentials, it can be shown that a digital computer obtains its impressive results solely by means of a sequence of complex switching operations. The computer contains a number of input terminals and a number of output terminals, on each of which an electric voltage may be either present or absent. In each switching operation voltage is applied to some input terminals of the computer and not to others, as a result of which voltage appears on some output terminals but not on others. The pattern of output voltage is rigidly determined by the pattern of applied input voltage, the internal wiring of the computer, and the electrical state in which each of its components has been left by any preceding switching episodes.

Indeed, if we ignore such practical matters as cost and size, it can be shown that any digital computer could be constructed in its entirety from a large number of electrically actuated switches. Each switch would be designed so that voltage would appear on its single output, or "power," terminal only on application of a suitable pattern of voltages to its several input, or "actuating," terminals.* A few of the actuating terminals would also double as computer inputs; similarly, a small fraction of the power terminals of the assemblage of switches would serve as computer outputs. But the vast majority of the interconnections within the computer would be governed by a very simple scheme: the power terminal of each switch would be connected to one of the actuating terminals of each of a number of other switches. To be sure, the details of such interconnections would have to be carefully worked out, in order to enforce the desired relationship between the sequence of output voltage patterns generated by the network and the sequence of input voltage patterns supplied to it. But this would be the only complication; the computer would still be properly described as consisting solely of multiply interconnected, electrically actuated, switches.†

---

* Some of the switches would operate, or fire, only on the appearance of voltage on a substantial fraction of the actuating terminals; others would be prevented from firing by such a condition. Still others would provide important memory functions, by the incorporation of features causing them to be affected by past conditions. For example, a time-delay element in its circuit could cause a switch to fire or not according to the actuating pattern on the *previous* occasion that it was energized. And an on or off state might be maintained indefinitely by another kind of switch, unless reset by the arrival of a suitable pattern of actuating voltages.

† Such a switch-only type of computer would also receive the raw material for its computations from switch devices. For example, one group of input switches might provide a coded representation of the pressure at some point in a chemical plant, by means of the pattern of voltages

Our interest, of course, is in whether there is anything in this basic switching nature of computers to justify ascribing to them relevance to the study of brain function. An affirmative answer is strongly suggested by the similarity between the essential switching-network structure of the basic computer and the anatomy of the brain. We have just seen that one can be described as a large number of multiply interconnected electric switches, arranged so that the output of each is determined by the pattern of impulses that comes to it from the other switches and input devices with which it is connected. But the other can be described as a large number of multiply interconnected neurons, arranged so that the output of each is determined by the pattern of impulses that comes to it from the other neurons and input sensory cells with which it is connected.* Such resemblance between the anatomy of a basic electronic computer and that of the nervous system would alone be enough to suggest a strong family relationship, even if there were no more fundamental reason to relate computers and the brain.

However, there *is* a more fundamental reason. It arises from a

---

formed by the on and off conditions of the members of the group. Similar input-switch groups might indicate pertinent values of temperature, flow rates, chemical composition, and so on. The periodic connection and disconnection of the power terminals of all such switches with the input terminals of the computer would then provide the required continuing sequence of input signals, to which the computer would respond by the generation of a suitably related sequence of output signals. These signals, in turn, appearing as patterns of voltages on the output terminals of the computer, would occasionally form coded constellations capable of activating external circuits to energize solenoids increasing or decreasing the consumption of power or flow of ingredients at various points in the system, or perhaps operating the keys of an electric printer to produce symbols or words meaningful to a human operator.

* And there is no scarcity, in the nervous system, of variety in the actuating requirements of the neurons similar to that displayed by the components of the man-made networks. In particular, some neurons are believed to possess history-dependent response characteristics adequate to form the basis of the important memory function.

certain powerful general capability that has been rigorously proved to be possessed by complex switching systems: if enough switching elements are used (and the number can be formidable), it is possible to interconnect them so as to enforce absolutely *any* detailed relationship between the sequence of voltage patterns appearing on the output terminals and the sequence previously applied to the input terminals, provided only that rules determining such a relationship can be expressed clearly and unambiguously.

But this is remarkable. In terms of our interests it can be paraphrased as follows: *If there is any kind of definite cause-and-effect relationship between the lifelong sequence of electric pulses leaving the brain and the lifelong sequence of pulses entering the brain, it can be precisely implemented by a switching network of the type that is known to underlie the design of all electronic digital computers and that at least appears to underlie the design of the brain.* It would be hard to imagine a discovery that would point more strongly both to the likelihood that the brain does achieve its results by purely physical means and to the probability that computers and brains are basically similar kinds of devices.

Before proceeding further along this exciting line of inference, we must observe that there is nothing here that requires anything like *identity* in the details of operation of brains and computers. Such a requirement would of course be untenable, even with reference to the basic, switch-only, type of computer. There are substantial differences among the electrochemical properties of the electronic and the neuronal components: a neuron is a much more complex device than a switch, and the nature of the signal propagated over its axon is by no means identical with the electric content of the wire that interconnects two switches. Indeed, most neurons rarely if ever operate as simple on/off switches. In computer terminology, a neuron possesses a combination of digital and analog characteristics: like a digital switching element, it does not ordinarily fire (except for an occasional random discharge) until it receives an adequate pattern of stimulation on its input terminals; nevertheless, when it does fire, the amount of stimulus it propagates to other elements over its long axon is not a fixed quantity, but depends on the magnitude of its own inputs.* There is also evidence that bulk chemical and electrical effects can modify the response characteristics of the neurons and thereby influence the overall operation of large regions of the brain.

But in terms of our interests, such complications are probably not very fundamental. It is true that the powerful general capabilities of switching networks have been derived by application of physical principles only to arrangements of the on/off components appearing in man-made digital computers. However, it seems certain that the same physical principles would attribute at least equal capabilities to the nervous system if the scientists were able to solve the much more difficult analytic problems that the greater operational sophistication of the natural components presents. To be sure, the optimization of the design of a network to take maximum advantage of the properties of the versatile neurons—and therefore the actual evolutionarily developed design of the brain—would be expected to involve details of interconnection of the components greatly different from those of existing computers. But again, although of enormous importance to those actually engaged in the design of improved computers, such practical differences between computers and brains do not greatly concern us. It is only necessary to remember that when henceforth we characterize the brain as a switching network, nothing so limited is implied as simple on/off switches interconnected like the components of an electronic digital computer of today.

* Although all voltage pulses propagated along the axon of a neuron are identical, a strongly stimulated neuron emits more such pulses per second than a weakly stimulated one, thereby producing larger effects in the downstream neurons or effector cells.

Experimental Test of the Computer/Brain Hypothesis

The chain of inference developed to this point runs about as follows: We set out in search of a purely physical explanation of intelligence. We observed that this required that the brain's output signals be causally related to its input signals through the operation of the ordinary laws of physics in the neuronal material. We then saw that the anatomy of the brain looks remarkably like that of the switching-network equivalent of electronic digital computers. We next learned of a powerful general theorem proving the ability of such switching networks to establish just the kind of cause-and-effect relationship between input and output that our initial assumption required. We therefore concluded not only that there is at least inferential support for the physical basis of brain function, but that also there is good reason to suspect that brains and computers employ similar operating principles.

But this argument clearly implies that it should be possible to design a computer with performance characteristics indistinguishable from those of a real brain. Such a computer, if provided with the same lifelong sequence of voltage pulses on its millions of input terminals as that brought to Einstein's brain over his incoming nerve fibers, would then provide at its millions of output terminals a lifelong sequence of voltage pulses indistinguishable from those which actually appeared in his motor-control nerves. Such a computer-generated sequence would, among other things, be able to direct the muscles of the fingers to write, and those of the throat to speak, all the details of the theory of relativity that Einstein did in fact write and speak during his lifetime.

It is likely that the construction and demonstration of a computer with such properties would convince nearly everyone that brains owe their impressive capabilities to the operation of physical principles essentially similar to those governing the performance of man-made electronic machines. However, severe practical problems stand in the way of such a convincing demonstration of computer intelligence. There are two main reasons why today's engineers could not be expected to build devices with properties approaching those of the natural articles. First there is the matter of *equipment* complexity. It is true that successive generations of machines have grown in complexity (and therefore in capability) until today hundreds of thousands or even a few millions of diodes, transistors, magnetic-core memory elements, and other electronic components may be employed in a single digital computer. But these numbers are still insignificant compared with the ten billion nerve cells each of us carries around inside his skull. In addition, as we have already observed, research has shown that each nerve cell is a device of considerable sophistication, when compared with the simple circuit elements of the computer designer.

The second impediment to the construction of brainlike computers arises from the *design* complexity of the problem. To specify the precise relationship between output and input that the general network theorem assumes to be known when it assures us that a corresponding computer can be designed, we would first have to solve some such problem as the following: "Given ten billion neurons with (presumed) known physical properties and (presumed) known interconnections, apply the known laws of physics to derive an expression describing the pattern of voltage found at any time on the millions of output neurons in terms of the preceding lifelong sequence of voltage patterns provided by the millions of input neurons." Even if we knew enough (which we don't) about the physical properties of the neurons and the details of their interconnections, the enormity of such a network problem places its rigorous solution well outside current capabilities.* Not even the remarkable enhancement of

---

* Not only does the human brain contain ten billion neurons, but on the average each neuron receives input connections from more than one thousand other neurons!

design power that can be achieved by putting large mathematically programmed digital computers to work on the task of designing still more complex computers is enough, today, to permit the problem to be solved.

Thus whatever may be the future potential of man-made computers, we must not expect too much of them now. Rather, we should bear in mind that their current inability to compete favorably with human brains in the performance of most intellectual chores may be indicative only of the very early stage of their evolutionary development. After all, the brain has had the advantage of a billion years of design improvement by nature's evolutionary processes; by comparison, the computer era is hardly more than twenty years old. While progress has been remarkable, it is probably not unreasonable for the engineers to ask for at least a hundred years or so to catch up with the advanced state of the art of such a competitor. Therefore the reader will be asked to be tolerant of some of the less than brilliant aspects of computer performance that we are shortly going to consider.

Now our next assignment is clear: it is to examine the efforts of computer scientists to design intelligent machines, and then to attempt to judge whether the subhuman nature of the observed performance seems adequately accounted for by the elementary nature of our understanding of the processes underlying intelligence and by the relative crudity of the presently available equipment. A favorable judgment, if we reach it, will lead to the conclusion that performance of the quality we call human will ultimately be possible for purely physical machines. The conviction this will lend to our belief in the essential identity of computers and brains will then support the thesis that the intelligence of animals and men must similarly be susceptible to an explanation in terms of the ordinary laws of physical science.

Let us now examine the evidence that computers, like men, are sometimes intelligent.

BIBLIOGRAPHY

Culbertson, J. T. "Some Uneconomical Robots," in *Automata Studies,* ed. by Shannon and McCarthy (Princeton University Press, Princeton, N.J., 1956), pp. 99–115.

Turing, A. M., "Computing Machinery and Intelligence," in *Computers and Thought,* ed. by E. Feigenbaum and J. Feldman (McGraw-Hill Book Company, New York, 1963), pp. 11–35.

chapter 9

The Intelligence of Computers

In this discussion our interest in computers is restricted to the light they can throw on the properties of living organisms. The case has already been made for the relevance of computer performance to brain function. Therefore we can move directly to the evidence for machine intelligence, without further discussion of how actual computers work.*

Computing by Computers

One thing computers can do is compute. Of all the branches of human logic, mathematics is easiest for electronics engineers to mechanize. By its nature it is rigidly formalized, employing symbols and rules of operation that can be precisely defined. And the

* Readers who are interested in a little more information about how modern computers actually accomplish some of the things about to be discussed will find it in the appendix at the end of this chapter.

logical operations involved are individually simple, lending themselves to easy simulation by sequences of switching operations on voltage patterns representing numbers. Because of this essential simplicity, together with the frequent appearance in human activity of problems requiring the performance of many successive mathematical steps for a solution, computations were the inevitable first assignments of the new electronic logic machines. This is why they came to be called computers. We should be thankful that their economic value to society in this limited area proved great enough to generate the large amounts of revenue needed to pay the salaries of the thousands of engineers and scientists whose work was necessary to develop their potentialities. Nevertheless, the computing capability of computers is probably the most nonbiological of all their attributes, holding the least interest for those who are concerned with how the brain may work. Despite its great economic importance, little will be said here about the purely mathematical facility of computers in the coming paragraphs.

Language Translation

"Modern guided-missile already possible carry with war head of hydrogen bomb and atomic bomb. Therefore it is one kind weapon with very big power of destruction." This was printed out by a general-purpose computer shortly after a human operator had typed into its input the characters expressing approximately same meaning in the Chinese language.* For practical reasons, a great deal of work has been done on the development of programs to cause general-purpose digital computers to translate from one language to another. In this country most

* As explained in the appendix to this chapter, general-purpose computers are assemblages of switching components whose internal connections are established through *programming* devices. We may think of programming such a computer as synonymous with wiring it up to enforce a desired input/output relationship.

progress has been recorded on translation from Russian to English. Chinese is more difficult but, as the above example shows, is not completely beyond the capability of modern machines.

Language translation is an activity that makes extensive use of the memory capacity of the computer. Obviously, some of the input information supplied to the translating machine must consist of the coded contents of a dictionary setting off words in one language against their equivalents in the other. In terms of our basic computer concept, it is pertinent to observe that such a dictionary could be built from pairs of switch groups, with one member of each pair carrying the electrically coded designation of an English word, the other member the coded designation of the same word in the other language. To be sure, when thousands of words must be stored in such an electronic dictionary, it is not really economical to use switches, so better components are employed; but the principle is the same.

Of course, intelligible translation involves much more than word-for-word substitution. The computer must be programmed to perform tests to determine the grammatical functions of the words in the input text. Subjects, predicates, direct and indirect objects, modifying phrases, and the like must be recognized for what they are and transformed into their corresponding meanings in English, frequently employing an entirely different word order. Idiomatic expressions and word endings characteristic of each language must be looked for and recognized. Ambiguities caused by different meanings for the same word must be resolved, usually by use of the syntactic information provided by the computer's analysis of the functions of the various words. Connective words and phrases, different in the two languages, must be suitably deleted or inserted. And so on.

Language translation is not easy for human beings, and it is not easy for computers. Machine translation is not yet as good as good human translation. Nevertheless, intelligible results are possible as a consequence of the ability of computers to perform such steps as those listed above. It is true that each step can be performed only because a clever programmer has been able to devise a set of mechanistic procedures to apply to the input symbols by means of which subjects, predicates, and the like can be routinely determined. Nevertheless, once the program is set into the machine, the know-how is there and does not have to be resupplied; any amount of text can then be automatically translated. It seems likely that a bilingual human being also unconsciously applies a set of standardized procedures in translating from one language to another.

Computers That Play Games

Proficiency in certain games has long been accepted by most people as indicative of at least a certain kind of intellectual capability. Checkers and chess would stand close to the top of most lists of games of this type. Digital-computer experts have made a number of attempts to program their electronic offspring to play such games.*

It might be thought that the high speed of electronic computers would make it feasible to program the modern chess- or checkers-playing machine to trace through to the end of the game all possible combinations of moves by itself and opponent and then unerringly to select the best one. This could reduce the computer play to a mechanical and unimaginative routine, im-

* The idea of a game-playing machine is not new, of course, and it must be admitted at the outset that modern electronic computers have as yet not equaled in performance a chess-playing machine constructed nearly two hundred years ago by Baron Wolfgang von Kempelen. This machine was described in detail by Edgar Allan Poe. For years it was taken on tour all over the world, defeating large numbers of excellent chess players including, it is said, Napoleon himself. The machine was also more compact than modern computers—so compact, in fact, that no one could believe there was a man inside it determining the plays and manipulating the levers until someone shouted "Fire!" during a game and caused the dwarf with the high IQ to struggle frantically to escape from his position of concealment.

pressive for its demonstration of electronic speed but irrelevant to the intellectual processes involved in human game playing. However, it turns out that such an approach is not feasible, even with the tremendous data-processing capacity of a modern digital computer. In a typical chess position, for example, there are 30 to 35 possible moves with 30 to 35 possible replies, giving a total of about 1,000 possibilities. An average chess game is 40 moves for each side. Thus, the total number of possible variations in a typical chess game is 1,000 multiplied by itself 40 times. If an attempt were made to program a computer to appraise all these possibilities, even though the computer could operate at the fantastic speed of 1 billion variations considered each second, the time required would still be impracticably large. In fact, if a game had been started between such a computer and the first cave man, not only would the machine not yet have decided on its first move, but it would still be only a small fraction of the way to its answer five or ten billion years from now, when the game would have to be called on account of the cooling off of the sun and the death of the solar system.

Thus there is no conceivable way of programming a machine to play a perfect game of chess or checkers in the sense of being able to examine all possibilities and inevitably decide on the best move. Instead, methods similar to those of human players must be employed. First a system must be devised for judging the favorableness of any configuration of pieces on the board, taking suitable account of the numbers, types, and positions of computer's and opponent's men. Then the condition of the board following various possible moves and countermoves must be examined to as great a depth as the machine capacity permits (only a few moves if a conclusion is to be developed in a few minutes' time). Finally the computer must select the move which, after these few plays and counterplays, leads to the configuration of pieces most favorable to the computer (assuming the human opponent also makes *his* best play each time, when evaluated according to the same rules).

A 1958 experiment in programming a digital computer to play chess was graphically described by Alex Bernstein and Michael de V. Roberts, of the IBM research laboratories.

You sit at the console of the machine with a chess board in front of you and press the start button. Within four seconds, a panel light labelled "program stop" lights up on the console and you now make your choice of black or white: to choose black, you flip a switch on the console; if you want white, you simply leave the switch as it is. Suppose you had picked black. To begin the game, you press the start button again. The machine now "thinks" about its first move. There is nothing spectacular about this. Some lights flash on the console, but the computer is working so swiftly that it is impossible to say just what these flashes mean. After about eight minutes, the computer prints out its move on a sheet of paper.

Let us say the machine's (white's) first move is king's pawn to the king's fourth square. The print-out then is W1 P-K4. The machine proceeds to print the chess board with the positions of the pieces, designating its own by the letter "M" and its opponents by the letter "O."

Now the "program stop" light goes on again, and the computer waits for its opponent to reply. You punch your replying move on an IBM card and put this card in a section of the machine which reads it. To signal that it is the machine's turn, you press the start button again. The machine prints your move and the new board position, and then goes on to calculate its second move. If you have made an illegal move, the computer will refuse to accept it, printing out "please check last move." So the game proceeds. At the end of the game, after a mating move or a resignation, the machine prints the score of the game, and to its opponent: "Thank you for an interesting game."

As to how good a game their machine played, Bernstein and Roberts reported:

In the first place, the machine is never absent-minded. It makes no blatant blunders, such as letting a piece be caught *en prise,* as every chess master has done at some time or other. When its opponent is careless enough to expose a piece, the machine takes instant advantage of the opportunity to capture it. Secondly, in its choice of individual moves, the machine often plays like a master, making what an expert would consider the only satisfactory move. Thirdly, the machine is certainly not in the master class in the play of a complete game. . . . yet notwithstanding its weaknesses, the IBM 704 plays a respectable and not too obvious game of chess—a game about which one can ask such questions as, "Why did it make that move?" and "What does it have in mind?" We can even say frequently that "it made an excellent move at this point," or, "At this stage, it had a good position." *

Since the Bernstein-Roberts work, considerable progress has been reported in the development of machine programs to make better game players of electronic computers. Probably the most outstanding accomplishment of this nature was that of A. L. Samuel of IBM, who succeeded in making of his machine a really championship-quality checkers player. To achieve this result, he went a long step further in equipping his computer with human problem-solving capabilities: he programmed it so that it could learn by experience and modify its behavior accordingly!

Despite its fundamental importance, it is not hard, in principle, to introduce machine learning into a computer program. In Samuel's checker-playing routine, the weight evaluation factors that were applied to the various aspects of the board configuration—numbers and types of pieces, positions, and so on—were made variable. When a move largely determined by the assumed favorableness of one of the immediately resulting configuration aspects led, later on, to a clear degradation in the computer's position relative to its opponent, the machine automatically reduced the value it subsequently assigned to that configuration aspect in its

* Reprinted from Alex Bernstein and Michael de V. Roberts, "Computer v. Chess-player," *Scientific American,* June, 1958, pp. 96–105. Copyright (1958) by Scientific American, Inc. All rights reserved.

future selection of moves; conversely, when experience showed that success usually followed moves that enhanced a certain feature of the board configuration, the machine automatically increased the value it subsequently assigned to that feature.

Samuel's checker-playing machine started out as a duffer, losing most of the games it played, even to inexpert opponents. But it learned. Indeed, it learned much faster than the average human player. After a few dozen games with a number of different opponents, the machine developed into a player of formidable capability. Consider, for example, the following report of Robert W. Nealey, the Connecticut checkers champion and one of the nation's foremost players, about a game he played against Samuel's computer in August, 1962:

> Our game . . . did have its points. Up to the 31st move, all of our play had been previously published, except where I evaded "the book" several times in a vain effort to throw the computer's timing off. At the 32–27 loser and onwards, all the play is original with us, so far as I have been able to find. It is very interesting to me to note that the computer had to make several star moves in order to get the win, and that I had several opportunities to draw otherwise. That is why I kept the game going. The machine, therefore, played a perfect ending without one misstep. In the matter of the end game, I have not had such competition from any human being since 1954, when I lost my last game.*

When performed by human beings, the translation of a foreign language or the playing of a game of skill is commonly considered to demonstrate intelligence. Is there a valid reason for employing any different word to characterize the machine performance of similar functions? Consider again Samuel's impressive checker-playing results. Can they not be described in biological terms as intelligent adaptation of behavior to environment?

* Reprinted from *IBM Research News,* August, 1962 (by permission of the publishers).

The machine observes and remembers the performances of itself and opponents; it abstracts from those performances the elements of significance; when enough experience has been accumulated, it generalizes to conclusions as to which of its own habits work best; finally it adapts its own behavior so as to improve its chances of achieving its goal—that is, of winning the games it plays. Prior to the development of modern computers, almost anyone presented with such a description would have said without hesitation that it represented intelligent behavior of the type exhibited only by higher animals. Of course, it could be that in programming a digital computer to learn and adapt its behavior from experience, we have only cleverly found a mechanical, non-thinking technique for doing something that we had erroneously believed required the exercise of intelligence. However, it is also possible that our own mental processes, when we understand them in detail, will similarly be found to reduce to a precisely ordered sequence of small-step, completely mechanical operations. This thesis will receive further support from the evidence of the next chapter that machines can be caused to display intellectual capabilities even higher than those we have so far considered, by the incorporation of operational features directly copied from the human nervous system.

APPENDIX *Operating Features of Existing Computers*

To begin with we should explicitly note that computers, like human beings, perform their higher intellectual chores in terms of symbols that stand for objects, actions, properties, or concepts. When used by human beings, systems of these symbols are known as language. Unsurprisingly, scientists and engineers speak of the language of a computer—the set of relationships between specific patterns of voltage applied to the input terminals and the resulting specific patterns of switching produced in the computer. As in the intercommunication of men of different nationalities, a translation operation is required if the computer and its human operator are to be able to talk to each other. This is accomplished by suitable devices at the input and output. For example, the operator may use an electric typewriter to feed data into the computer. Hitting the 2 key of the typewriter then operates switches that set up the pattern of voltages on the computer's input terminals that it recognizes as a 2. Similarly, some of the output terminals can be connected to a printer in such a way that specific keys will be automatically operated in accordance with the generated patterns of voltage, thereby translating the electrical language of the computer into the written language of the human operator. And letters as well as numbers can be fed into the input or extracted from the output. Of course, the human operator will be able to make sense out of the written product of the computer's activities only if its internal construction provides for switching operations that cause the sequence of output symbols to have some kind of logical relationship to the sequence of input symbols. Improperly designed or faulty computers, like infants and mentally disturbed adults, may speak only nonsense.

Another important feature of most modern computers is their general-purpose nature. Although it is necessary for the internal connections among the components to be set up correctly to enable a desired relationship to be established between input and output sequences, computer scientists have devised an ingenious method of establishing and modifying such internal connections so that the same physical equipment can be used in a large number of different applications. Over and above the switching elements that are needed in the actual implementation of a desired input/output relationship, additional electrically operated switches are arranged so that, by energizing them selectively, different modes of interconnection of the first set of elements can be established. Thus the computer can in effect be instantaneously "rewired" to solve a new problem—that is, to set up a new form of relationship between sequences of input and output patterns.

The ease with which the internal wiring can be changed by the method just described leads to an important economic principle in computer design: it is best to build the machine so that it can, through such switching, establish a large number of relatively simple patterns of relationship between input and output, rather than a small number of complex patterns. For the computer scientists have been able to show that if the basic repertoire of computer modes is properly chosen, any complex relationship between sequences of input and output signals can be established by successive application of the simpler permitted transformations. Therefore, in the operation of such a general-purpose computer, *each* of the successive patterns of voltage comprising the input sequence triggers a bewildering series of related switching steps: First the input pattern activates a relatively simple network that, by several successive switching operations, produces a short subsequence of output voltage patterns. Some of these output patterns are stored in sets of electrically operated switches or the equivalent (the computer's memory) for later use. The switches controlling the internal wiring of the components then are caused to shift to a new pattern of on-ness and off-ness, thereby setting up the computer to establish a new relationship between input and output. In the new mode, the stored results from the previous steps, together with the current computer input pattern, activate the new network to generate a new subsequence of output patterns, some of which again are stored in the memory. Again the internal wiring is changed by operation of the controlling switches; again a new subsequence of intermediate outputs is generated; and so on. In a typical general-purpose computer hundreds of such small transformation steps may be employed to generate the same final output voltage pattern that the basic, permanently wired computer could generate in a single massive switching operation. But because the electronic components operate so quickly (frequently in much less than one-millionth of a second), such long series of simple steps still permit problem solutions to be obtained in reasonable periods of time.* And the practical advantages are usually substantial, because of the resulting equipment economy and versatility.

The step-by-step switching of the internal wiring of the components is controlled by a specific computer *program*. For each step,

---

* The great speed of operation of the electronic components partially compensates for their small number, compared with the neurons of the brain. But only partially—after full allowance is made for such compensating factors, the modern computer is still a trivial piece of equipment compared with the human brain.

the electrically operated switches determining the details of the internal connections are selectively energized by means of a suitable pattern of voltages placed on a group of special input terminals. To illustrate the principle, we may imagine a human operator establishing the control pattern of voltages by typing symbols on an electric typewriter each time the computer is ready to perform the next step in its series of transformations. However, since modern computers work so much faster than any human typist, such instructions are generally prerecorded (on magnetic type, say) and then fed into the computer fast enough to keep up with its operation.

It is, of course, the precise nature of the series of small input/output switching sub-sequences controlled by the program that determines whether the task performed by the computer is the preparation of an income tax return or the determination of the next move in a game of chess. Most of the subject matter of this chapter consists of evidence that some of the programs that can be devised for general-purpose computers cause them to perform chores which, when performed by human beings, are called intelligent.

A reaffirmation of the pertinence of computer performance to the capabilities of the brain may be in order. The employment by most actual computers of tape-controlled programs, with the resulting appearance in their operation of large numbers of successive small transformation steps instead of massive switching episodes, does not in any way invalidate the general principles discussed in Chapter 8. These features are straightforward consequences of design expediency related to the currently available components and techniques and the economic advantages of general-purpose machines. Every computer operation discussed *could* be performed by the more apparently biological kind of machine consisting of nothing but permanently interconnected electrically actuated switches.

BIBLIOGRAPHY

Bernstein, A., and M. de V. Roberts, "Computer v. Chess-player," *Scientific American,* June, 1958, pp. 96–105.

Garvin, P. L., "A Linguist's View of Language-data Processing," in *Natural Language and the Computer,* ed. by P. L. Garvin (McGraw-Hill Book Company, New York, 1963), pp. 109–127.

King, G. W., and H. W. Chang, "Machine Translation of Chinese," *Scientific American,* June, 1963, pp. 124–135.

Mersel, J., and coworker, "Machine Translation Studies of Semantic Techniques," report by Thompson Ramo Wooldridge Inc. to ARDC, USAF, on work done under contract AF 30(602)-2036, Feb. 22, 1961.

Samuel, A. L., "Some Studies in Machine Learning Using the Game of Checkers," in *Computers and Thought,* ed. by E. Feigenbaum and J. Feldman (McGraw-Hill Book Company, New York, 1963), pp. 71–105.

Simon, H. A., "How Computers Can Learn from Experience," in *Applications of Digital Computers,* ed. by Freiberger and Prager (Ginn and Company, Boston, 1963), pp. 11–27.

chapter 10

Machines That Imitate the Brain

Pattern Recognition

One of the conspicuously nonhuman attributes of the game-playing computers of the last chapter was their dependence on the careful ministrations of a human operator to supply, in machine language, suitable indications of the moves made by the opponent. Could a computer be equipped with a lifelike visual system enabling it to look at the board and learn where the pieces are, just as we do?

Of course, the issue at question is much more profound than this limited statement of the problem might suggest, for it has to do with the ability of higher animals and human beings to extract meaningful information from the kaleidoscopic and ever-changing patterns of light and shade focused on the retinas of their eyes. This in turn involves neuronal and intellectual processes of great subtlety, and is by no means thoroughly understood today. This incompleteness of our understanding of perception must prevent any very complete demonstration of its mechanical nature. Nevertheless, progress has been made: machines have been equipped with at least a rudimentary quality of visual perception.

To be able to see, the machine must of course first be provided with the equivalent of an eye. This poses no great problem to the electronics engineer. Usually a flat surface is studded with a number of small photoelectric cells, so that when a pattern of light and shade is focused by a lens on this surface, there results a related pattern of electric currents in the bundle of wires leaving the cells. The analogy between the photoelectric cells and the rods and cones of the retina of the eye is of course obvious, as is that between the bundle of output wires and the optic nerve whose millions of individual axons carry visual information to the brain. And the machine/nervous-system analogy can be extended even further, for in the machine the wires from the photoelectric cells are connected to the actuating terminals of a number of electric switches, just as are the axons of the optic nerve connected to the "actuating terminals" of neurons in the brain.

Suppose that the task we want our machine to perform in order to prove its possession of a basic perceptual ability is that of identifying properly each of the letters of the alphabet (a disappointingly simple chore by human standards, perhaps, but we must always remember that our machines are still in a very early phase of development). So that we human beings can follow its performance, we may equip the machine with twenty-six output switches (effector, or motor, neurons, as it were), one for each letter, and arrange that the firing of each output switch is signaled by the illumination of an associated indicator light. Correct operation of the machine will then consist in the fact that the B and only the B light goes on when the letter B is shone on the retina of photocells, and so on.

Now a circuit with the properties specified to this point could be designed as easily and with as little pertinence to the processes of intelligence as the animated electric signs in Times Square.

For by using enough switches, any electronics engineer could work out a set of interconnections among photoelectric cells and output indicators to achieve the desired result. But we have one more condition to add: the circuit must not be prewired in this way to perform its assigned function; instead, it must be capable of modifying its internal structure as a result of experience—in short, it must be able to learn to make the required identifications. Now, it is clear, we have specified a kind of machine performance with at least some resemblance to that of a biological perceptual system.

It turns out that there is an interesting way to implement this important new requirement. In its simplest equipment form, the output wire of each photoelectric cell of the retina of the machine is connected through an adjustable resistance to an actuating terminal of *each* of the twenty-six output switches.* The effect of each of these interconnections in encouraging its switch to fire depends on the value of the interposed resistance. Changes in these values constitute the modifications of internal structure by means of which the circuit learns.

Training involves the successive focusing on the retina of different illuminated letters. The presentation of each letter results in the appearance of voltage on some of the actuating terminals of each output switch. As a result, some of the switches fire, others do not. At this point an auxiliary teaching circuit goes to work. If an output switch has fired when it should not have done so, the teaching circuit turns the switch off again by increasing the series resistances associated with the effective actuating terminals—those which happen to be receiving voltage from the illuminated retina. Conversely, an output switch which should have fired but didn't is urged to action by a decrease of the resistances in the voltage-carrying actuating inputs. In this way the

---

* Actual pattern-recognition circuits are usually more complex than this. They are also frequently able to perform more sophisticated discriminations.

circuit is gradually trained. By repetitive presentation of the twenty-six input patterns and successive small selective increases and decreases in the input resistances, a circuit balance is finally achieved wherein the presentation of the letter A at the retina always results in the illumination of the A indicator light only, and so on.

An interesting incidental property of the circuit is that it can be taught to identify letters correctly even when they are imperfectly drawn: there is a certain tolerance, due to the learning process, in the required precision of illumination of the retina. And by the supplementation of the simple equipment so far described with other banks of electrically actuated switches (the machine's interneurons) that sum the effects of the individual photoelectric cells in various ways before connecting them to the output switches, circuits can be built which can be trained to identify handwritten characters of different sizes and shapes.

Of course, our interest in this kind of demonstration is proportional to its biological relevance. We have already adequately noted the similarities between electronic aggregations of multiply interconnected switches and nervous-system aggregations of multiply interconnected neurons. But there is an even more suggestive aspect to these electronic pattern-recognition networks, related to the teaching circuit which controls the making of successive small increases and decreases in the strengths of interconnection of the circuit elements. For substantially all theories of biological learning postulate that a remarkably similar mechanism goes to work in an organism whenever a sensory stimulus results in a "correct" or an "incorrect" response—that is, one leading to pleasure or one leading to pain. In such a training episode it is believed that the pleasantness or unpleasantness of the experience triggers some sort of electrochemical reaction which strengthens or weakens the interneuronal connections, or *synapses,* in the currently firing stimulus/response network, and therefore makes it more or less likely that a future repetition of the same stimulus will lead to the same response.

Thus an equipment technique deliberately selected to resemble one of the more conspicuous dynamic features of nervous systems results in machines that can learn to make simple visual discriminations. Few scientists believe this to be mere coincidence.

Problem-solving Machines

Probably we would all agree that most of our intellectual activity consists of the solution of the problems we encounter from day to day. "How can I see the ball game and still get my work done?" "How can I buy a new car and keep my bills paid?" "How can I increase my income?" "How can I pass the car ahead of me without risking an accident?" Or, "How can I design an undercarriage that will soft-land a spaceship on the moon?" It is in his ability to find solutions to such problems that man's superiority to other animals resides. It would be of the greatest significance if, by applying knowledge gained from a study of human problem solving, machines could be successfully programmed to perform such chores. Impressive progress has already been made in this direction.

Of course, although we may hope that ultimately machines will be able to deal with all the nuances of such complex concepts as "ball game," "income," and "spaceship," it is impracticable to require such an intellectual capacity of currently available models. Thus we must allow the computer scientist certain liberties in divesting any proposed problem of the rich conceptual embellishments with which human language ordinarily surrounds it before posing it to his machine for solution. Today an adequately impressive demonstration would consist of the programming of a computer just to perform the same *kinds* of logical operations that human beings go through when they attempt to solve difficult problems, even though these operations might have to be carried out in a somewhat abstract setting.

Fortunately, mathematicians and logicians have made good progress in identifying the fundamental steps that underlie most human problem solving. We can, for example, replace one concept by its logically equivalent one—the work we have to get done before going to the ball game can be transformed into the three hours of time that it requires. We can also perform an operation to reduce the difference between two things we are trying to equate—by the operation of deferring until tomorrow one hour's report writing, we can reconcile the time required for doing our work with the two-hour interval before the start of the game, and so on.

Whether we are deciding about going to a ball game or devising a new theory of relativity, we appear to use only a handful of such basic logical processes. It has therefore been possible to formalize the essence of logical thought in a discipline called *symbolic logic*. In this discipline objects or propositions are represented by letters or other convenient symbols, and other symbolic notation is employed to specify the different kinds of operations on, or transformations of, these objects or propositions that the rules of logic will allow. With such symbolic notation, the logical steps actually employed by human beings in their solution of real problems can be concisely written down and studied.

It was this reducibility of the essence of human logic to symbolic representation that first suggested the possible applicability of computers, with their symbol-manipulating abilities, to a broad range of problems requiring the performance of logical processes. As long as ten years ago this possibility was under serious investigation. In 1957, for example, three of the pioneers in the field, A. Newell, J. C. Shaw, and H. Simon, of the RAND Corporation and Carnegie Institute of Technology, reported on early results with their Logic Theory Machine. They described the object of this work as "to learn how it is possible to solve difficult problems such as proving mathematical theorems, discovering scientific laws from data, playing chess, or understanding the meaning of English prose."

The sophistication of the problem-solving techniques brought to bear by machines equipped with the kinds of programs that have been developed in recent years can be indicated by a recital of a few of the steps a machine may take when presented with such a problem as the proof of a geometrical theorem.* To solve such a problem, the machine, like a human student, must make use of the basic axioms and previously derived theorems of Euclidean geometry. Therefore these, as well as the statement of the problem to be solved, are stored in the machine memory, in suitably coded form. The machine may then start by a routine application of each of the axioms and given theorems to the various elements of the geometrical figure under consideration, to learn whether such a simple approach will prove the theorem in question. When it does not, the machine program will cause it to choose a *subgoal* to work on. The simplest subgoal may be a condition or theorem which, if proved, will then permit the attainment of the *goal*—proof of the final theorem—by the application of one of the axioms or given theorems. Human problem solvers frequently choose such subgoals by working backward from the desired solution; they say to themselves, "What simpler theorem can I work on which, if proved, will then lead to the proof of the main theorem?" Sometimes the subgoal chosen by a good geometer will lead to the final solution only through a fairly complex chain of logic, involving not only the successive application of several axioms and given theorems but even the use of discoveries about the particular geometrical configuration being worked with which have been made in the course of previous unsuccessful attempts at solution of the problem. The machine chooses subgoals and works on them in the same way. Every time an axiom or given theorem is applied to some part of the geometrical configuration under study, the resulting new relationship is stored in the machine memory as part of the body of axioms and valid theorems for use in later steps of the problem-solving effort. Thus, with the passage of time, the machine learns how to apply more and more powerful tools to the problem at hand, choosing

---

* Much of the work on proving geometrical theorems was done at the IBM Research Center in New York, by H. Gelernter and associates.

subgoals that are related to the final goal by more and more complex chains of logic.

Each subgoal, while it is under investigation, is a full-blown problem in its own right. Thus the machine, like the human problem solver, may choose and investigate subgoals for the proof of a higher subgoal, just as it chooses and investigates subgoals for the proof of the goal itself.

In working backward from the desired solution, the human problem solver must usually exercise judgment in his selection of subgoals. An unimaginative approach would lead to a routine listing of all possible statements, or subgoals, which, if true, would imply the final proposition, followed by a similar listing, for *each* of these subgoals, of the corresponding set of statements, or sub-subgoals, capable of proving the subgoal, and so on. The indiscriminate following down of each such possible line of attack is usually out of the question for the human problem solver. Indeed, it is also out of the question for computers; despite their great speed of symbol manipulation, it has been estimated that the discovery of proofs in this way for geometry problems of average high school difficulty would frequently require thousands of years of computer time! Thus the machine problem solver, like the human problem solver, must be able to employ judgment to select the more promising lines of attack. In the solution of problems in geometry, one way of making such judgments involves the use of a diagram. Frequently most of the subgoals for a higher-order goal can be quickly eliminated by such means. For even though the problem under consideration could easily be solved if it could first be established that line segments AB and CD were equal, for example, there would be no point in wasting time on such an approach if it were obvious from a diagram drawn in accordance with the stated problem conditions that AB and CD were *not* equal. At least one successful geometry-problem–solving computer program employed such a test in selecting which lines of attack to follow.

Of course, working backward from the desired solution is not the only technique used by human beings in solving problems. Sometimes, instead, the method is to make some kind of judgment of the difference between what is known and what needs to be proved, then to look for logically permitted transformations or operations that might reduce the difference. The human problem solver may, for example, have proved that two line segments are equal, whereas he knows that the theorem he must prove is expressed in terms of angles; thus he may feel that the application of axioms and valid theorems to convert his line-segment relationship into another relationship involving angles will give him a better chance of discovering a solution to the goal or subgoal under consideration. Problem-solving machine programs have been devised that also make use of such loosely defined measures of relatedness of different kinds of expressions and undertake transformations to minimize the observed differences.

Much more could be said about the programming of electronic computers to solve problems in logic, but what has been said should suffice to illustrate the important points. Permitted transformations are performed on initial conditions in an attempt to reach the solution of the problem; when this is impossible, subgoals are generated that are capable of leading to a solution; sub-subgoals are generated that are capable of establishing subgoals; choices are made of the more promising subgoals and sub-subgoals to work on by the application of certain kinds of judgment processes; sometimes the degree of relatedness of different expressions is sensed, and operations are performed to bring them closer together; learning continually goes on to increase the sophistication of successive solution attempts. Clearly such processes justify the application of the adjective "intelligent." Yet the machine functions entirely automatically: once the human operator feeds into its memory the statement of the problem together with the axioms and given theorems, he has nothing to do but fold his arms and wait until the electric typewriter, seconds or minutes later, types out the steps of the final proof (or, sometimes, a confession of inability to solve the problem).

How skillful are such machines? In proving theorems in plane geometry, performance about equivalent to that of a good high school student has been demonstrated: sometimes the machine is faster, sometimes slower; sometimes the machine fails on problems considered simple by the student, other times it develops elegant and ingenious solutions to very difficult problems. Similarly, when programmed for the solution of problems in symbolic integration, machines have demonstrated performance comparable to that of the average college freshman. And machines have also demonstrated significant competence in general symbolic logic, having derived proofs for most of the fifty or sixty basic theorems that beginning students of the subject are required to prove. In all this work the machines have been programmed to employ problem-solving techniques similar to those commonly used by human beings.

Indeed, it is important to keep in mind this biological orientation of our recent discussion. In this chapter we have added a condition to our earlier specification of the kind of research results we will consider interestingly pertinent to an understanding of how the brain works. Here we have not been considering just *any* evidence for the existence of machine intelligence, but instead have been focusing our attention on intelligent behavior arising out of the application of general operational principles that seem similar to those we employ in our own thinking. Thus we have attributed significance to the ability of one machine program, using a generally anthropomorphic problem-solving method, to derive most of the fifty-odd elementary theorems of symbolic logic, while we have chosen to ignore the accomplishment of another machine in proving all 350 of the logic theorems in Russell and Whitehead's *Principia Mathematica* in less than nine minutes. This is because the latter accomplishment involved programming principles that, while making effective use of the mathematical competence of the machine, appear to have little relation to how human beings solve problems.

From this narrowly man-centered point of view, the most pertinent of all demonstrations would consist in the programming of a machine, not just to display generally manlike intelligence, but to duplicate the actual thought processes of a specific individual. Even this ambitious objective has been approached by several groups of investigators.

Machine Simulation of Intellectual Processes of Specific Individuals

A. Newell and H. A. Simon, already referred to as pioneers in the study of machine problem-solving techniques, have compared the detailed intellectual approaches of machines and human individuals to the solution of the same problems in symbolic logic. Their work was based on machine programs designed to duplicate as closely as possible the methods of thought their studies had indicated were employed by their human subjects. After being equipped with such a program, the machine was then caused to type out a running account of the various steps it tried in working on a new problem, and a similar account was obtained from the human problem solver by requiring him to think out loud—to describe his reasoning as he tried various symbolic-logic processes in his attempt to solve the same unfamiliar problem. On comparing the machine and human recitals, numerous substantially identical reasoning sequences were discovered. Occasionally the machine and the human problem solver would make different choices of a next subgoal or step, but on the whole there was a remarkable degree of similarity between the paths of human and machine trial and error on the way to the problem solution. Newell and Simon concluded that this work revealed "with great clarity that the free behavior of a reasonably intelligent human can be understood as the product of a complex but finite and determinate set of laws."

Some successes have even been reported in computer simulation of the intellectual activity of individuals engaged in the practical decision making required by nontechnical professions. For example, one investigator, G. P. E. Clarkson, undertook to mechanize the judgment processes employed by a particular trust officer in a bank in deciding what common stocks to buy for his clients. In addition to accumulating general rules used by trust officers in determining the suitability of stocks for investors, the computer scientist obtained from the officer he was studying a set of statements of his reasons for past decisions, as well as comments on the ideas, forecasts, and facts presented in various financial-journal articles and analysts' reports. On the basis of such information a computer program was devised which, it was hoped, would apply to the various kinds of data bearing on the selection of common stocks about the same weight factors that the trust officer in question had been applying in his past decisions. The resulting machine program was then presented with current stock-exchange data on a large number of stocks, together with pertinent information about four of the bank's new clients, and was required to predict the portfolio the trust officer would select for each. For the smallest account the trust officer selected five stocks, with thirty, fifty, ten, fifty, and fifty shares respectively; the computer selected a different first stock, but predicted the other four selections, with forecast purchases of fifty, ten, sixty, and forty-five shares respectively. On another account seven stocks were actually purchased; of these, the computer predicted six correctly, including the number of shares. The third account involved eight stocks, seven of which were chosen accurately by the computer. In the largest account the computer missed two stocks out of nine.

Computer programs have even been designed to predict decisions of the members of the United States Supreme Court. In the analysis of a particular class of cases—those pertaining to right to counsel—Reed C. Lawlor has shown the internal consistency and predictability of the votes of the individual justices. His computer was able to score 86 percent accuracy in the prediction of individual votes, making use of a program based on the voting records of the same justices on other cases.

The growing acceptance of the idea that man's mental processes are ultimately explainable in terms of the regular operation of natural law was interestingly demonstrated in June, 1962, when prominent psychologists and computer scientists collaborated in a symposium at Princeton University on the topic "Computer Simulation of Personality." The work discussed involved the programming of electronic computers not just to perform processes we ordinarily call intellectual but also to exhibit what would be considered, in human beings, emotional and sometimes even neurotic responses to the environment.

Rate of Evolution of Computer Intelligence

As we approach the conclusion of this discussion, it may be in order to warn against overoptimistic inferences as to the probable future rate of evolution of computer intelligence. For in this kind of treatment there is a danger of making it all appear too easy. A reader unfamiliar with computers could conclude from this review of their accomplishments that only a few years may be needed to bring the IQ of man-made machines up to a human level. This would be incorrect. Current progress is not that fast.

Indeed, computer experts are inclined to be less sanguine than they once were about quick success in the development of machines and techniques suitable for a steadily widening range of sophisticated intellectual assignments. This is only in part a reaction to some of the relatively uninformed optimism of a few years ago. In addition, the logical processes involved in thinking are proving to be harder to understand and mechanize than many had expected. As a result, the rate of progress in machine simulation of intelligent activities, initially rapid, has at least temporarily leveled off. It is not clear that computer translation of foreign languages is much better now than it was five years ago; the world's championship in chess is still held by a human being, not a machine; the field of automatic pattern recognition has recorded few recent advances. Hence more is heard now than formerly of the limitations of computers—for example, that in the near future they should not be expected to replace human intelligence, but only to supplement it by routine data-processing assistance.

We can easily sympathize with the computer scientists who sometimes appear more impressed by the present difficulties than by the long-range prospects of their field. For the likelihood that twenty or thirty years will not suffice to bring intelligence theory and machine design to a point where human capabilities can be equaled must be of the greatest importance to those whose work is coupled to the practical products that flow out of it. To us, however, primarily concerned as we are with the validity of the inference that the goal will *ultimately* be achieved, the time scale of progress is of secondary interest. There is no reason why the problems should be easy. The fact that they are proving to be difficult does not weaken the case we have examined for the fundamental similarity of computers and brains.

BIBLIOGRAPHY

Clarkson, G. P. E., "A Model of the Trust Investment Process," in *Computers and Thought,* ed. by E. Feigenbaum and J. Feldman (McGraw-Hill Book Company, New York, 1963), pp. 347–371.

Gelernter, H., "Realization of a Geometry-theorem Proving Machine," in *Computers and Thought,* ed. by E. Feigenbaum and J. Feldman (McGraw-Hill Book Company, New York, 1963), pp. 134–152.

Gelernter, H., J. R. Hansen, and D. W. Loveland, "Empirical Exploration of the Geometry-theorem Proving Machine," in *Computers and Thought,* ed. by E. Feigenbaum and J. Feldman (McGraw-Hill Book Company, New York, 1963), pp. 153–163.

Hawkins, J. K., "Self-organizing Systems: A Review and Commentary," *Proceedings of the Institute of Radio Engineers,* 1961, pp. 31–48.

Lawlor, R. C., "What Computers Can Do: Analysis and Prediction of Judicial Decisions," *American Bar Association Journal,* April, 1963, pp. 337–344.

Mattson, R. L., "A Self-organizing Binary System," *Proceedings of the Eastern Joint Computer Conference,* 1959, pp. 212–217.

Newell, A., J. C. Shaw, and H. Simon, "Empirical Explorations with the Logic Theory Machine," *Proceedings of the Western Joint Computer Conference,* 1957, pp. 218–239.

Newell, A., and H. A. Simon, "GPS, a Program That Simulates Human Thought," in *Computers and Thought,* ed. by E. Feigenbaum and J. Feldman (McGraw-Hill Book Company, New York, 1963), pp. 279–293.

Ridgway, W. C., III, "An Adaptive Logic System with Generalizing Properties," Technical Report no. 1556-1, prepared under Air Force contract AF 33(616)-7726, April, 1962.

Selfridge, O. G., and U. Neisser, "Pattern Recognition by Machine," in *Computers and Thought,* ed. by E. Feigenbaum and J. Feldman, (McGraw-Hill Book Company, New York, 1963), pp. 237–250.

Tomkins, S. S., and S. Messick, *Computer Simulation of Personality* (John Wiley & Sons, Inc., New York, 1963).

Uhr, L., and C. Vossler, "A Pattern-Recognition Program That Generates, Evaluates, and Adjusts Its Own Operators," in *Computers and Thought,* ed. by E. Feigenbaum and J. Feldman (McGraw-Hill Book Company, New York, 1963), pp. 251–268.

chapter 11

Summary of the Argument

The argument of these chapters on intelligence constitutes an important part of the modern explanation of the properties of living organisms. It deserves summarization and restatement.

At the beginning of Chapter 8, on computers and the brain, we asserted that an adequate explanation of human activity we call intelligent would require establishing that it is completely and in detail the consequence of the operation of the ordinary laws of physical science in the material of the brain cells, under the continuing stimulus provided by the lifelong sequence of incoming neuronal signals. But in following the modern attempts to prove this thesis, we did not encounter the straightforward approach of applying the known laws of physics to the neuronal network and thereby deriving intelligence as an inevitable result. The brain is too complex and the understanding of what constitutes intelligence is too primitive for such an approach to be productive at this time. Therefore we found ourselves considering an indirect method of probing the mysteries of the intelligent brain—the study of the properties of electronic computers.

Of course, the history of science provides many precedents for the solution of complex problems by extrapolation from the solutions of simpler ones. However, when this approach is used, the relevance of the simpler to the more complex phenomena becomes a matter of key importance. In our discussion, therefore, attention was paid to the strength of the case that can be made for the conclusion that electronic computers and human brains employ similar principles of construction and operation.

We found, first, a resemblance between the underlying design principle of the computer and the most conspicuous anatomic feature of the brain. We learned that one can be described as a large number of multiply interconnected electric switches, arranged so that the output of each is determined by the pattern of impulses that comes to it from the other switches and input devices with which it is connected. And we saw that the other can be similarly described as a large number of multiply interconnected neurons, arranged so that the output of each is determined by the pattern of impulses that comes to it from the other neurons and input sensory cells with which it is connected.

We found, next, that computer scientists had succeeded in establishing that any kind of definite cause-and-effect relationship between the lifelong sequence of electric pulses leaving the brain and the lifelong sequence of electric pulses entering the brain could, in principle, be precisely implemented by a switching network of the type that is known to underlie the design of all electronic digital computers and that at least appears to underlie the design of the brain. It was concluded that it would be hard to imagine a discovery that would point more strongly both to the likelihood that the brain does achieve its results by purely physical means and to the probability that computers and brains are basically similar kinds of devices.

But if such inferences were indeed correct, it was clear that computers should be capable of intelligent performance—of a low order compared with that of the human brain, perhaps, owing to the relative crudity of existing equipment, but nonetheless performance in which the essential ingredients of intelligence could be identified. In effect, we believed such a demonstration of machine intelligence was held to constitute a critical test of the validity of the physical theory of human intelligence.

Accordingly, we then undertook a review of some of the evidence that computers can indeed be caused to perform chores which, when done by men, are considered to require the intervention of intelligence. We found that machines have demonstrated substantial capabilities in the translation of languages, the playing of games of skill, the recognition of handwriting, and the solution of problems in symbolic logic. In performing these tasks the machines were seen to "choose" goals and subgoals, perform "logical" operations on objects or propositions, and "judge" the degree of relatedness of different expressions. They were seen to "remember" information provided to them, as well as the results of their own logical activities. They were observed to "generalize" on the results of past experience, to "deduce" valid conclusions therefrom, and then "adapt" their own future behavior accordingly. In general, an ability to "learn from experience" was found to be a widespread property of computers. Thus the existing evidence showed that many of the terms we use to describe our own mental processes are also applicable to the operations of modern machines. Indeed, we learned that machines can be designed not only to employ generally human methods in solving problems but also to reproduce some of the detailed thought sequences of specific individuals.

It was finally decided that such evidence constituted a powerful argument for the conclusion that *all intelligence, whether of computer or brain, is the natural consequence of the powerful symbol-manipulating capabilities of complex switching networks and that therefore the ordinary laws of the physical scientist are adequate to account for all aspects of what we consider to be intelligent behavior.*

Probably this part of this book should end, as it began, with a warning against allowing subjective feelings about the nature of thought to obscure the logic of the evidence. In particular, it must not be imagined that reduction of the processes of intelligence to small-step mechanical operations is incompatible with the apparently spontaneous appearance of new and original ideas to which we apply such terms as "inspiration," "insight," or "creativity." To be sure, there is no way for the physical methods we have been dealing with to produce full-blown thoughts or ideas from out of the blue. But it will be recalled that there is a solution for this problem. The solution is to deny that such spontaneity really exists. The argument is that this is an example of our being led astray by attributing too much reality to our subjective feelings—that the explanation of the apparent freedom of thought is the incompleteness of our consciousness and our resulting lack of awareness of the tortuous and detailed, step-by-step nature of our thought processes.

With this observation, we come face-to-face again with the problem of consciousness and of how it can be explained in purely physical terms. The reader has suspended his skepticism on this point long enough. We proceed next to a consideration of this important and difficult subject.

part 4

Consciousness

chapter 12

Consciousness—
Important Though Passive

In Chapter 7 we tentatively adopted two working hypotheses about consciousness. It was there asserted that these hypotheses would later on be found consistent with the conclusions to be developed in our considerations of intelligence. We should verify this consistency before going on to a more extensive study of consciousness.

One of the postulates—the unconscious nature of some of our mental activity—requires little justification. The competent regulation by the nervous system of complex metabolism during sleep or anesthetization demonstrates that consciousness is not a necessary concomitant of many neuronal processes. Other observations are even more pertinent. For example, a mother can sleep through a wide variety of loud neighborhood noises, yet, through unconsciously acting sensory discrimination mechanism, be roused to instant alertness by the faint cry of her child. And a cat that has been trained to recognize one musical tone as heralding an unpleasant electric shock and another of only slightly different

133

pitch as innocuous will be awakened by the one but not by the other. These seemingly intelligent judgments may be performed as skillfully when we are unconscious as when we are conscious; indeed, a cat's ability to discriminate between similar tones has been found to be better when it is asleep than when it is awake. Because of such observations, it is not hard to accept the related conclusion that, in general, many of the details of our thought processes are hidden from conscious view.

Our second hypothesis proclaimed the passivity of consciousness. The hypothesis meant simply that physical science alone should be capable of providing an accurate and complete explanation of all the externally observable properties and actions of organisms, including not only those we call physical but also those we call behavioral. The critical test of the hypothesis was conceived to be whether or not intelligent behavior is susceptible to a purely physical explanation. The material of the last five chapters has supported such a conclusion.

It is important that the implications of this concept of passivity be clearly understood. The ability of purely physical processes to account for *all* externally observable details of behavior means, among other things, that no aspect of another person's activity—tone of voice, facial expression, appearance of the eyes, or any of the other cues by which we judge what is "in his mind"—can provide proof that he experiences the kind of awareness that we call consciousness. By the same token, every detail of the past and future history of mankind would be the same if consciousness were completely nonexistent, just so long as the physical laws of nature were kept unchanged.

This concept of passivity, which is widely held by life scientists, makes it possible for the investigator of behavior and intelligence to ignore consciousness entirely, if he chooses to do so, with confidence that such a deliberate narrowing of subject matter will not invalidate any conclusions he may reach. However, the resulting scarcity of references to consciousness in many papers on the mechanisms of behavior and intelligence does not mean that the scientists involved consider it nonexistent or unimportant. Even though it may be irrelevant to the control of behavior, each of us knows that consciousness is a real phenomenon; none would deny his own possession of it. And to us it certainly *seems* important—perhaps more important, indeed, than the "objective" properties of the universe that scientists usually deal with. Therefore, despite the difficulty of the subject, consciousness has not been completely ignored by research workers. Some investigations have been specifically designed to probe selected aspects of conscious phenomena; pertinent discoveries have also been made as by-products of other kinds of brain research. In the course of this work much evidence has been uncovered that while consciousness may not exercise influence over the physical processes in the material of the brain, these processes can and do exercise great influence over consciousness.

Indeed, in the next two chapters we are going to examine evidence for the thesis that both the existence and the content of consciousness are completely determined by the detailed electrochemical state of material in the brain. We shall then consider the implications of such a thesis with respect to the eligibility of the properties of consciousness for inclusion among the phenomena governed by the regular and predictable operation of a single body of natural laws.

chapter 13

The Physical Source
of Consciousness

We know that most of the parts of the body have little or nothing to do with consciousness. Arms and legs can be lost without interference with the continuity of content of the conscious state that causes a person to be the same individual today that he was yesterday. The same is true of internal organs such as lungs and kidneys, as long as their decay or removal is compensated by medical techniques so as to sustain life. Even the heart, designated by the ancients as the seat of the soul, has been shown to be irrelevant to consciousness by recent surgical achievements involving extensive modifications of the natural organ. To be sure, at this writing the complete replacement of a human heart by a man-made device has not yet occurred, but certainly no one expects that such operations, when they are possible, will in any way compromise the integrity of personality of the patients.

Thus, even if there were no more direct evidence, the process of elimination would focus attention on the brain as the locale of whatever physical activities are required for the appearance of the conscious state. And without analyzing it deeply, we all do indeed associate the brain with the personality and individuality. When the Russians succeed in grafting a second head onto a dog, we have little doubt that two canine personalities still exist and that, despite the absence of a body of its own, the transplanted head is somehow aware of being the same dog it always was. And when Dr. White and other Western Reserve University scientists succeed in keeping a monkey brain alive after it has been removed from the animal, we are inclined to believe that their electric indications of alert brain activity do indeed mean that the organ is conscious, despite its disembodied state.

The Seat of Consciousness in the Brain

Even if we accept the general conclusion that consciousness is a result of some kind of activity in the brain, it is still meaningful to ask, "Just where in the brain is the seat of consciousness?" For the brain is a most complex organ. It is in fact a collection of organs; anatomists and physiologists employ dozens of names to refer to the various parts of the brain. However, for our purposes it will be sufficient to speak of three general regions: the *brainstem, cerebellum,* and *cerebral cortex.*

The brainstem was briefly mentioned in connection with our discussion of reflex activity; it is the deepest-lying part of the brain, being really an extension and elaboration of the spinal cord. It is sometimes called the *old brain,* inasmuch as the evolutionary developments that have made of man a modern and advanced animal have not had a great effect on this part of his nervous system. The brainstems of frogs, lizards, pigeons, rabbits, monkeys, and men look much alike.

The cerebellum, in man, is a baseball-sized lump of neuronal tissue protruding from the back of the brainstem. It varies greatly in appearance from animal to animal, but in a rather special way. Relatively enormous in birds and almost nonexistent in lizards,

the cerebellum is known to be involved in the control of locomotion and balance.

The cerebral cortex is the part of the brain that appears to be
most related to intelligence and to have been most affected by
evolution. A swelling scarcely larger than other brainstem pro-
tuberances in the frog, the cerebral material becomes steadily
larger as we move up the scale of animal intelligence. As this material grows from one species to the next it pushes backward, ex-
panding around the brainstem until, in monkeys, apes, and men,
the cerebral material by its growth has become a covering (that
is, cortex) that surrounds and hides most of the other regions of
the brain. This ⅙-inch-thick sheet of gray-colored tissue is draped
around the other parts in such a way as to fill tightly all the re-
maining space inside the skull. The wrinkles, folds, and convolu-
tions of the human cortex give the impression that nature has
gone to considerable pains to pack into the limited space available as much yardage of this sheet material as possible.

Because the cerebral cortex appears to be associated with intelligence and we are inclined to feel there is a connection between
intelligence and consciousness, we might expect this part of the
brain to be most directly related to the presence or absence of the
conscious state. And indeed there is some evidence that this is the
case. For example, *decorticate* animals—cats or dogs whose cortex
has either been removed or disconnected—can live indefinitely
with their autonomic physical functions relatively unimpaired;
but they appear to be asleep or unconscious.*

On the other hand, not all the cortex is necessary to sustain
consciousness. Sickness and accident have provided many in-

* As already noted, if the evidence for the passivity of consciousness is
accepted, it is impossible for us to know whether any other person or
animal possesses the sense of personal awareness we know *we* have. Here
and subsequently we will use a commonsense, rather than a logically
rigorous, approach: we know our own state of consciousness is paralleled
by certain physical manifestations of alertness; therefore we assume the
presence or absence of similar physical manifestations to be indicators of
the presence or absence of consciousness in other persons or animals.

stances of men and women who remained effective members of society after the loss of much of their cortical tissue. Indeed, people have been found capable of many years of intelligent, competent, and alert behavior after the destruction or removal of one entire cerebral hemisphere. (The brain is symmetrical in design, like much of the body; the two similar halves of the cortex are called the left and right cerebral hemisphere.)

Thus these observations suggest that the existence of what we know as the state of consciousness requires the physical presence and healthy condition of a substantial part, though not all, of the cortical material. While we are dealing here with inference rather than proof, let us tentatively accept the conclusion and see where this line of thought leads us.

The Consciousness Switch

It is clear that the presence and health of cortical material, while perhaps necessary, is certainly not sufficient to guarantee the existence of the conscious state. For the healthiest of animals go to sleep from time to time. And all can be rendered unconscious by general anesthesia. Apparently consciousness requires that the right kind of electrochemical activity be under way in the cortex. Something is now known about the nature of this activity. For a "consciousness switch" has been discovered in the brain.

Near the top of the brainstem, or old brain, there is a mass of neurons that has been given the name *reticular activating system* by H. W. Magoun, the physiologist at the University of California at Los Angeles who is largely responsible for the work we are about to discuss. Nature appears to have gone to great pains to cause essentially all the incoming and outgoing communication channels of the brain to pass through the reticular system. This is done largely by means of *collaterals*. For example, a nerve carrying sensory information from the spinal cord to the cortex does not go directly through the reticular formation, but as it passes by, its main fibers send off smaller branches to terminate on reticular neurons. A collateral arrangement is also found in the motor nerves as they pass by the reticular formation on their way from the higher centers of the brain to the main cable of the spinal cord. And in addition to these wire taps on the nearby communication channels, the reticular activating system also has direct lines of command to the stations of interest to it. These receiving stations include half a dozen major areas of the cortex and probably all the nuclei of the brainstem.

The extensiveness of its interconnection with other parts of the brain suggests an important role for the reticular activating system. And indeed Magoun and coworkers found it to be the source of regulating signals controlling many of the unconscious response patterns of the nervous system. But they also found it to exercise control over the property of consciousness itself.

For example, stimulation of the upper part of the reticular activating system by an electric current injected through an implanted electrode* induces sleep in cats and dogs, provided that the stimulating voltage is caused to wax and wane with a long, slow period. Of even greater significance is the fact that stimulation with a rapidly changing wave form always awakens the sleeping animal.

To be sure, the arousal of a sleeping animal by direct electric stimulation of the brain is not in itself impressive, since wakefulness can be elicited in so many ways—by a touch, a noise, a bright light, and so on. The significance arises from experiments showing that the *only* thing that arouses a sleeping animal is electrical activity originating in the reticular activating system— that other sensory stimuli affect consciousness solely through the collateral currents they send into the reticular formation, and not by means of their direct connections to other parts of the brain.

---

* Tiny wires can be surgically inserted into any part of the brain of man or animal and led out through holes in the skull, without damage to the patient. Human beings have worn such attachments for more than two years without ill effects.

For example, lesions confined to the reticular activating system of cats and monkeys will induce permanent coma; lesions interrupting the long sensory pathways from the body to the higher centers of the brain but sparing the reticular activating system permit a continuation of the wakeful state. It is now known that general anesthesia produces its effects by deactivating the neurons of the reticular system. A pinch on the foot, a sound in the ear, or a light in the eye will produce electric sensory currents in the cortex of the brain that are as strong and clear when the subject is anesthesized as when he is conscious and alert.* The patient is unaware of these sensations only because under the influence of the drug, the reticular system is unable to generate and send to the cortex the additional special pattern of electric signals needed to turn on the state of consciousness. Of course, the same absence of alerting input also inactivates the automatic muscular reflexes that might otherwise be elicited by the sensory experiences of the operating table.

While the reticular activating system acts as a switch to turn consciousness on or off by sending suitable electric signals to the parts of the brain involved in conscious processes, ordinarily it is itself urged into action by the sensory impulses it receives over its wire taps on the communication channels of the central nervous system. Incoming signals, including those representing touch, pain, sound, or light, are integrated by the reticular neurons to build up an output voltage to a value high enough to cause the reticular system to send out its arousal commands. In the absence of such incoming sensory data, direct electric stimulation of the reticular formation by implanted electrodes will fool the mechanism into believing that there is something that requires conscious attention.†

* These electrical effects are directly measured by means of wires inserted into the cortex and connected to electrical instruments.

† Even though consciousness itself may have no direct influence on the capabilities of the organism, the improved response properties that happen to accompany consciousness can be well worth arousing.

One particularly interesting kind of nervous activity employed by the reticular neurons in the summing process leading to the arousal signal is that which goes on in the cortex. The reticular formation gives such activity as high a weight as external sensory stimuli in its determination of the need for consciousness. This is why, whenever we are aroused from sleep by a loud noise, we remain awake for some time after the noise has ceased. The thought processes set in motion on our awakening create cortical currents which, by the collaterals into the reticular activating system, provide a continuing stimulus that keeps the consciousness switch turned on. As we all know, occasionally to our sorrow, thinking alone can keep us awake. If, however, we can sufficiently diminish the intensity of our intellectual activity, the cortical currents alone may no longer be able to hold the reticular consciousness switch closed. Therefore, unless there remain reticular-activating-system currents caused by noise, pain, light, or other sensory stimulus, the consciousness switch opens, and we fall asleep.

The Major Conclusion

While some of the evidence connecting the cerebral cortex with consciousness may not by itself be completely convincing, that related to the consciousness switch is. Here there is nothing vague or indeterminate. A distinctive signal is generated in the brainstem when sensory stimulation or cortical activity is high enough. When this signal is able to reach the cortex, the subject is conscious. When it is interrupted, whether due to a sleep situation in the normal brain, anesthesia, disease, or surgical section of the communicating fibers, the subject is always unconscious.

Here is incontrovertible proof of a regular and predictable relationship between physical activity in brain tissue and the presence or absence of consciousness. In the next chapter we shall examine evidence that the content of consciousness, as well as its existence, may be determined by physical processes in neuronal material.

BIBLIOGRAPHY

French, J. D., "The Reticular Formation," *Journal of Neurosurgery,* vol. 15 (1958), pp. 97–115.

Hernandez-Peon, R., H. Scherrer, and M. Jouvet, "Modification of Electrical Activity in Cochlear Nucleus during 'Attention' in Unanesthetized Cats," *Science,* vol. 123 (1956), pp. 331–332.

Magoun, H. W., *The Waking Brain* (Charles C Thomas, Publisher, Springfield, Ill., 1958).

Wooldridge, D. E., *The Machinery of the Brain* (McGraw-Hill Book Company, New York, 1963), chap. 7, "Control Centers of Emotion and Consciousness."

chapter 14

Physical Determinants
of the Content
# of Consciousness

Sensation

Sensation is an important ingredient of the conscious state, and much is known about the physical events which produce it. For example, there is a specific region in the cortex within which electrical activity must occur if we are to feel something touching our skin. Called the *sensory strip,* this cortical material serves as a receiving station for signals coming in through the nerves from the body surface. Pinching a finger will result in electric current in one part of the sensory strip; mistreating a toe will result in current in another part of this cortical region. And if a wire is inserted into the exposed sensory strip of a conscious human patient and electric current is injected from an external generator, the patient will report a tingling sensation in the part of the body that is connected by sensory nerves to the cortical spot being

144

stimulated.* For it is the electrical activity in the sensory cortex that is interpreted by the consciousness mechanisms as stimulation of the body surface. When electrical activity is artifically supplied, the rest of the brain is fooled into thinking that something is going on at the periphery of the body, even though this may not be true at all.

The part of the brain involved in consciousness can be fooled in other ways than by artificial stimulation of the exposed cortex. Severing or otherwise mistreating the nerves leading to the sensory strip can result in patterns of electrical activity in the cortex that achieve a similar effect. This is why an amputee can feel the fingers of his hand being twisted in an uncomfortable position for months after the arm carrying that hand has been cut off.

In addition to the sensory cortex, there is an *auditory cortex,* where electric stimulation produces in the patient a sensation of clanging noise. There is also a *visual cortex,* at the very back of the head. In one part of this region, stimulation might result in the sensation of a flashing light in the lower left of the field of view; in another part, it might result in the appearance of the fireworks in the right center of the visual field.

Such sensations as hunger and thirst have also been found to be caused by electric currents in specific small regions of the brain. While these regions, or *centers,* are in the brainstem rather than the cortex, their properties also display an intimate relationship between the content of consciousness and localized neuronal activity. For example, when stimulating electric impulses are sent through a surgically implanted wire into the appropriate part of an animal's *appetite center,* the animal will exhibit the only kind of appetite that truly deserves the adjective "insatiable:" it will continue to eat any food provided to it, even though it is so stuffed that it must regurgitate what it has already swallowed. If

---

* Electric stimulation in human beings is done in connection with brain surgery, not research. We will learn more about the use of this technique a little farther on.

the stimulating electrode is placed in another part of the appetite center, only a fraction of an inch away, the animal will be unable to eat, even to the extent of starving to death while surrounded by its favorite food.

These sensory regions of the brain are the same for all individuals of a given species. No learning is involved; the nervous connections are wired in at birth. They are determined by the genetic specifications, just as are the number and shape of arms and legs.

The conclusion strongly indicated by this kind of evidence is that the sensational content of consciousness is always and in detail the result of purely physical causes—the genetically determined wiring diagram of the nervous system and the physical activity in the neurons. The mystic's vision may indeed come to him without any corresponding pattern of light and shade on the retinas of his eyes, but it probably does require a corresponding pattern of electrical activity among the neurons that control the visual content of consciousness.

Emotion

There are qualities of the conscious state to which we apply such terms as "feeling" and "emotion." A feeling or emotion is more general, less localized, than the sensations we have been discussing. Also it *seems* like a different kind of thing. Evidence that emotions, like sensations, are the direct results of specific patterns of electrochemical activity in the brain would be particularly helpful in establishing a case for the physical basis of conscious phenomena. And there is such evidence. For centers have been discovered in the brainstem where electric stimulation produces intense emotional effects.

In a monkey, for example, stimulation within a suitable small region of brainstem tissue causes the animal to display the manifestations of rage: its hair stands on end, it growls and bares its teeth, it claws or bites anything or anyone who comes near. Injection of electricity into another part of the brain seems to produce fear: the animal wants to run and hide. Sheer horror appears to be the state produced by stimulation in other of the *punishment centers* of the brain.

There are *pleasure centers* as well as punishment centers. When a wire electrode is implanted in such a center in a rat, for example, and a sort of telegraph key is provided whereby the animal can self-administer pulses of electricity to its brain, it will quickly learn to pound away at the key thousands of times an hour. In such experiments animals have continued such incessant self-stimulation for twenty-four to forty-eight hours, stopping only when physical exhaustion made further operation impossible. And on being put back in the cage after a few hours of sleep, with food in one corner and the key in the other, the hungry animal has been observed to ignore the food, dash to the key, and indulge in another orgy of self-stimulation!

Pleasure and punishment centers were early found in the brains of rats, chickens, cats, dogs, and monkeys. Of course, these are animals, not human beings, and some have questioned the relevance of animal experiments to consciousness, holding that all that can be observed is *apparent* emotional response. Therefore it is pertinent to note that in connection with the treatment of patients suffering with one form or another of brain-related abnormality, in recent years the same kinds of centers have been discovered in the human brain. These patients have confirmed that stimulation in suitable sites causes feelings of horror, rage, fear, ease, relaxation, or ecstasy.

In short, there is now extensive evidence that the emotional, as well as the sensational, content of consciousness is directly determined by the pattern of electrochemical activity in specific regions of neuronal tissue.

Conceptual Thought

We pass now to evidence that the intellectual content of consciousness, as well as its sensational and emotional content, is determined by the detailed physical state of neuronal material.

Some pertinent observations have been made by brain surgeons. The pioneer in this field is Dr. Wilder Penfield, for thirty years head of the Montreal Neurological Institute. Penfield's discoveries came in connection with his use on human patients of the technique of electric stimulation of brain tissue, which he brought out of the animal-research laboratory and converted into a useful surgical tool.

Penfield's technique was employed in operations to remove tumors or epileptically defective brain tissue. It worked about as follows. After removal of the section of skull covering the area where defective tissue was believed to lie, the patient would be restored to consciousness in the operating room, and asked to describe his sensations as needles were stuck into various parts of his brain and electricity injected. (The brain contains no nerve endings that can sense pain; a patient suffers no physical discomfort in such tests.) By the nature of the responses, it was frequently possible to determine the boundaries of defective tissue more accurately than by visual observation alone.

One important by-product of this new surgical technique was the discovery that a patient's ability to speak can be impaired by electric stimulation in certain cortical regions. This would not have seemed particularly unusual if the effect produced had just consisted of interference with the operation of the muscles of the mouth and throat. But Penfield's stimulation seemed to interfere with thought processes rather than with muscles. For the patients had no difficulty in talking per se; they were simply unable to think of certain key words as long as the electricity was being injected. On being asked to name the object in a picture while the speech area of his cortex was being stimulated, one patient said: "Oh, I know what it is. That is what you put in your shoes." After withdrawal of the electrode he said, "Foot." A little later he was unable to name the picture of a tree, although he knew what it was, naming it properly as soon as the stimulus was turned off. Another patient under electric stimulation could not think of the word for a comb, although when asked its use he said, "I comb my hair." When asked again to name it, he couldn't until the electrode was removed.

From descriptions by the patients of their sensations when these tests are under way, it is clear that what is involved is interference with only a part of the speech mechanism. The patient recognizes the object he is trying to name, can speak effectively in most respects, but for some reason no longer has access to some common words that ordinarily would come to him easily. He frequently tries to think of a synonym when he finds he cannot remember a word he needs. On one occasion, a patient struggled unsuccessfully to answer a question during the electric discharge; afterward he said, "I couldn't get that word 'butterfly' and then I tried to get the word 'moth.' " The test is evidently a puzzling and frustrating experience for the subject.

While these results were produced by the artificial injection of electric current into the exposed cortex of the patient, the behavioral effects were quite similar to those observed in the same and other patients when the stimulating current was injected into the speech cortex as a result of spontaneous epileptic attacks.* The principal difference was that natural causes seemed to generate a wider variety of symptoms. For example, lesions occurring in the part of the speech cortex that abuts the visual cortical region were sometimes found to produce much more severe disturbance of writing than of speech. Other cases were observed in which the patient could express himself adequately but could not understand what was said to him. Such cases, however, are comparatively rare. Usually the general capability for symbolic representation of thought seems to be interfered with. A patient with an extensive region of defective tissue in the speech cortex not only is unable to express himself in speech but is similarly handicapped in conveying meaning by gesture of head or hand. He may use the muscles of the neck for other purposes, but he can-

---

* Epilepsy is known to produce its bizarre effects by a discharge of excess electricity from defective tissue into surrounding healthy regions of the brain.

not nod assent in place of the word "yes" or shake his head in place of the word "no." Gestures are lost as well as words; they too are symbols of conceptual thought.

Memory

A second important by-product of the use of the electric-probe technique in brain surgery was the discovery of certain effects related to memory. In 1936, several years after the first use of his new technique, Penfield was preparing to operate on a woman whose lesion was on the side of the cortex just above the ear—a region that later work was to implicate with the memory mechanisms. When the needle electrode was inserted and the electric current turned on, the patient suddenly reported that she felt transported back to her early childhood. In the operating room, she essentially relived an episode out of her remote past, even feeling again the same fear that had accompanied the original event.

In the succeeding years, Penfield and other brain surgeons have observed, in many patients, this kind of triggering of memory by cortical stimulation. The phenomenon possesses a number of fascinating features, which the developing understanding of brain function must some day explain. For example: The electrically elicited experiences always appear to be reproductions of real events although the patient usually has not been consciously carrying them in his memory. The induced recollection can be stopped abruptly by turning off the stimulating current and can often be restarted by turning it on again. But when restarted, the recalled episode does not continue where it left off. Instead, it starts again from the beginning, as though it were stored on a film or tape which automatically rewinds each time it is interrupted!

Perhaps most startling is the vividness of the elicited experience. Instead of being a remembering, according to Penfield, "it is a hearing-again and seeing-again—a living-through-moments-of-past-time." Nevertheless, the patient does not lose contact with the present. He seems to have concurrent existences—one in the operating room and one in the part of the past that he is reliving. The name *double consciousness* is employed by brain surgeons to describe these peculiar sensations of their patients.

Splitting One Individual into Two

Penfield's elicitation of the phenomenon of double consciousness was a particularly startling discovery. The idea that the single individual that each of us feels himself to be could ever be split into any semblance of two separate individuals seems incomprehensible; nothing could be subjectively harder for us to accept. And nothing could more strongly imply that the content of consciousness is determined by the physical state of brain tissue. It is therefore of great importance that Penfield's discovery no longer stands as the sole example of successful tampering with the unity of consciousness. In the Caltech biology laboratories, Dr. R. W. Sperry has had under way for several years a series of experiments that in some ways are even more spectacular.

Sperry's methods are based on the left/right symmetry that characterizes higher animals, including cats, monkeys, and men. In particular, as already indicated, in such animals the brain consists of two similar hemispheres. They are interconnected by a vast system of nerve fibers numbering, in a human being, several hundred million.

In concept, Sperry's idea was simple: break these connections and see what happens. The trouble was that when he tried it on laboratory animals (usually cats), nothing much did happen. Even after most of the connections between the brain hemispheres had been cut, the animal appeared substantially unchanged. It behaved about as before; it achieved about the same scores on intelligence tests; it remembered its old tricks and could learn new ones. The evidence suggested that the two halves of the normal brain are essentially carbon copies of one another, each containing a complete complement of memories and control mechanisms.

The experiment had to be refined. This was done by adding to the brain-splitting surgery an operation on the visual system of the animal, as a result of which the right eye was left connected only with the right half of the brain, and the left eye with the left half. (Ordinarily there are also crossing connections; these were cut.)

Cats prepared in this way were then subjected to special training procedures. They were taught to choose between two swinging doors that they could easily open, one carrying a conspicuous circular design, the other a cross.

The cat under training would learn that choice of the door with a circle, say, would lead to a reward; choice of the door with a cross would lead to punishment. This was routine animal training; the distinguishing feature of the experiment was that the cat was provided with an eyepatch, so that all of its trials involved the use of, say, the left eye only, which was connected to the isolated left half of the brain.

The animals learned their lessons about as fast as cats that had not had their brains split. Everything seemed normal—up to a point. This point was reached when the cat had been trained to near perfection using one eye and then was reintroduced into the training cage with the eyepatch shifted, so that the untrained eye —and the associated half of the brain—came into play.

The result was spectacular. Changing the eyepatch was exactly equivalent to changing cats. When employing the eye connected to the untrained half of its brain, the animal appeared to have not the slightest recollection of ever having been in the problem box. It could be trained again, using the new eye and hemisphere, to perform the desired discrimination, but its rate of learning was exactly the same as that of an entirely fresh, untrained cat.

In the new condition, the cat could even be taught, equally easily, a discrimination opposite to that learned originally. Using its right eye and right brain hemisphere, it could learn to open the door with the cross despite its earlier acquisition, when using its left eye and left hemisphere, of the habit of opening the door with the circle. Such a doubly trained animal would shift its performance automatically, without confusion, when the eyepatch was shifted.

In some respects, the split-consciousness implications of the work on cats was demonstrated even more vividly by later experiments on monkeys. In this work the two halves of the brain were provided with different properties, not by training, but by surgical modification. On one side of the monkey's brain, the nerve fibers running from the brainstem to a forward part of the brain were severed. This operation, a *frontal lobotomy,* has for years been known to produce significant personality changes if performed on both sides of the brain. Specifically, it produces under these circumstances a relaxed, "I don't care" person or animal. Before the development of tranquilizing drugs, it was sometimes employed to ease the unbearable tensions of psychotic mental patients.

In addition to the frontal lobotomy on one side of the brain, the monkeys were subjected to the split-brain operation, including modification of their optic connections, just as in the cat experiments. After the surgery, each was fitted with an arrangement of contact lenses similar in effect to the cats' eyepatches.

A monkey employing the eye that was connected to the unmodified half of its brain was then shown a snake. Monkeys are normally deathly afraid of snakes, and a split-brain monkey was no exception: it showed the usual fright and escape reactions. Then the conditions were changed so that the monkey had to use the eye connected with the hemisphere that had had the lobotomy. Again the snake was displayed. This time the monkey could not have cared less; the snake held no terrors for it. It was as though two different animal personalities now inhabited the body that had formerly been occupied by one.

Despite the spectacular nature of this conclusion, there can no longer be much doubt of its validity, for similar and even more convincing observations have now been made on human beings. In 1962, at the White Memorial Hospital in Los Angeles, the split-brain operation was performed on two severely handicapped epileptic patients. It was hoped that cutting most of the connections between the two sides of the brain might decrease the severity of their massive epileptic seizures. Happily the operations were both unexpectedly successful, in that they almost eliminated the attacks completely. But our interest is in the postoperative psychological testing on the patients, which was conducted in the Caltech laboratories. These tests revealed clearly that splitting the brain results in splitting the personality in human beings as well as in cats and monkeys.

As Sperry has pointed out, this was illustrated in many ways.

For example: the subject may be blindfolded and some familiar object such as a pencil, a cigarette, a comb, or a coin placed in the left hand. Under these conditions, the mute * hemisphere connected to the left hand feeling the object perceives and appears to know quite well what the object is. Though it cannot express this knowledge in speech or in writing, it can manipulate the object correctly, it can demonstrate how the object is supposed to be used, and it can remember the object and go out and retrieve it with the same hand from among an array of other objects either by touch or by sight. While all this is going on, the other hemisphere meanwhile has no conception of what the object is and, if asked, says so. If pressed for an answer, the speech hemisphere can only resort to pure guesswork. This remains the case just so long as the blindfold is kept in place and all other avenues of sensory input from the object to the talking hemisphere are blocked. But let the right hand cross over and touch the test object in the left hand; or let the object itself touch the face or head as in demonstrating the use of a comb, a cigarette, or glasses; or let the object make some give-away sound, like the jingle of a key

---

* In these patients, as in normal human beings, speech was controlled by the *left* hemisphere of the cortex. The left hand is, however, connected to the *right* hemisphere. *D. W.*

case, then immediately the speech hemisphere also comes across with the correct answer.*

Other tests showed that the personality in the left half of the brain had no awareness of visual images coming from the left and therefore projecting to the right cortical hemisphere, and that the right-hand personality was similarly oblivious to what was being seen by the left. And although the participation of the unsplit brainstem in the control of gross body movements seemed usually to prevent extensive conflict between the physical activities of the two sides of the body, there were even some instances in which the left and right hands, apparently motivated by opposing objectives, worked at cross purposes—one attempting to put on a garment, for example, with the other simultaneously attempting to take it off.

From tests and observations made over the course of two years, Sperry concluded that "what is experienced in the right hemisphere seems to be entirely outside the realm of awareness of the left hemisphere. This mental division has been demonstrated in regard to perception, cognition, volition, learning, and memory." In short, "Everything we have seen so far indicates that the surgery has left these people with two separate minds, that is, two separate spheres of consciousness." †

The Major Conclusion

The properties of the sensory cortex, the existence of pleasure and punishment centers in the brain, the electrical interference with the speech mechanisms, the peculiar double consciousness of Penfield's patients, the split personality of Sperry's split-brain subjects—all are variations on a single theme: the operation of the

* Reprinted from R. W. Sperry, "Brain Bisection and Mechanisms of Consciousness," in *Brain and Conscious Experience,* ed. by J. C. Eccles (Springer-Verlag New York Inc., New York, 1966), pp. 299–300 (by permission of the publishers).

† *Ibid.,* p. 299.

principle of physical cause and effect in the content and quality of the conscious state.

We feel pain or cold, hunger or thirst, horror or ecstasy, *because* of the electrochemical activity in specific regions of brain tissue; we appear to have concurrent existences in past and present *because* of the disturbance of our brain mechanisms caused by Penfield's electric probe; we have two separate states of consciousness *because* Sperry has cut the connections between the two halves of our brain.

The inference is a strong one: the content of consciousness, as well as its presence or absence, is determined in detail by the physical structure and electrochemical state of the material of the brain. This conclusion has far-reaching implications. We shall examine some of them in the next chapter.

BIBLIOGRAPHY

Buchanan, A. R., *Functional Neuro-anatomy* (ed. 4, Lea & Febiger, Philadelphia, 1961), chap. 12, "The Sensory and Associative Mechanisms of the Cerebral Cortex."

Burns, B. D., *The Mammalian Cerebral Cortex* (Edward Arnold [Publishers] Ltd., London, 1958), chap. V, "The Problem of Memory."

Hess, W. R., *The Functional Organization of the Diencephalon* (Grune & Stratton, Inc., New York, 1957).

Olds, J., "Differentiation of Reward Systems in the Brain by Self-stimulation Technics," in *Electrical Studies on the Unanesthetized Brain,* ed. by Ramey and O'Doherty (Harper & Row, Publishers, Incorporated, New York, 1960), pp. 17–51.

Olds, J., "Pleasure Centers in the Brain," *Scientific American,* October, 1956, pp. 105–116.

Penfield, W., and L. Roberts, *Speech and Brain Mechanisms* (Princeton University Press, Princeton, N.J., 1959), chap. II, "Functional Organization of the Human Brain, Discriminative Sensation, Voluntary Movement," and chap. III, "The Recording of Consciousness and the Function of Interpretive Cortex."

Sem-Jacobsen, C. W., and A. Torkildsen, "Depth Recording and Electrical Stimulation in the Human Brain," in *Electrical Studies on the Unanesthetized Brain,* ed. by Ramey and O'Doherty (Harper & Row, Publishers, Incorporated, New York, 1960), pp. 275–290.

Sperry, R. W., "Brain Bisection and Mechanisms of Consciousness," in *Brain and Conscious Experience,* ed. by J. C. Eccles (Springer-Verlag New York Inc., New York, 1966), pp. 298–313.

Sperry, R. W., "The Eye and the Brain," *Scientific American,* May, 1956, pp. 48-52.

Sperry, R. W., "The Growth of Nerve Circuits," *Scientific American,* November, 1959, pp. 68–75.

Trevarthen, C. B., "Double Visual Learning in Split-brain Monkeys," *Science,* vol. 136 (1962), pp. 258–259.

Wooldridge, D. E., *The Machinery of the Brain* (McGraw-Hill Book Company, New York, 1963), chap. 2, "The 'Schematic Diagram' of the Nervous System"; chap. 7, "Control Centers of Emotion and Consciousness"; chap. 8, "Personality and Speech"; and chap. 9, "Memory."

chapter 15

Transfer of Consciousness from Metaphysics to Physics

We have learned that determination of the conscious effects caused by alterations of structure or electrochemical activity in specific regions of the brain lies within the capability of the experimental-research scientist. The orderliness and reproducibility of the results have supported the conclusion that the content of consciousness, as well as its presence or absence, must be determined in detail by such physical structure and activity. In principle, at least, we can contemplate the ultimate knowability of the precise conscious consequences of any given pattern of interconnection and stimulation of the brain. Such lawful and predictable conduct would appear to qualify consciousness as a scientific subject. Indeed, as has been implied in Chapter 7, if we wish to use "physics" to refer to a single body of laws and particles underlying biology, chemistry, and all other natural sciences, we must find a way to expand its definition to include the basic principles related to conscious phenomena.

158

The nonscientist may feel that there is something unnatural about a suggestion that consciousness is suitable for absorption into the subject matter of physics. It may seem to him that even though lawful relations between the content of consciousness and the physical * state of matter are eventually established, this will not really explain consciousness: we still won't know what it is or where it comes from. And this is true—nothing yet proposed provides any hope of explaining consciousness in a fundamental way. Further, it is safe to predict that no such explanation ever will be discovered. But precisely the same statement could be made about mass, electricity, or gravity.

The scientist may invent a term such as "gravitational attraction" to enable him to discuss a phenomenon he wishes to deal with, and he may agree with other scientists on techniques for measuring the attraction between material objects. He may then perform experiments that ultimately permit him to deduce laws, or relations, connecting gravitational forces and the masses and positions of the bodies involved. Thus he can learn how to predict the gravitational effects that will be produced by a specified configuration of objects or, conversely, how to arrive at valid configurational deductions in terms of measured gravitational forces. But for questions like "What *is* gravity, really? What causes it?" and "Where does gravity come from? How did it get started?" the scientist has no answers. Long familiarity with the regularity and predictability of gravitational effects may cause him to lose the sense of mystery which once surrounded the subject. Nevertheless, in a fundamental sense, it is still as mysterious and inexplicable as it ever was, and it seems destined to remain so. Science can never tell us why the natural laws exist or where the matter that started the universe came from. Thus the fact that consciousness also possesses an essential mystery and inexpli-

---

* Despite the possible incorporation of the phenomena of consciousness in the subject matter of physics, common practice dictates limitation of the use of the adjective "physical" to nonconscious properties of matter.

cability does not weaken its claim of eligibility for inclusion in the subject matter of physics.

Of course, none of this can invalidate the possession by consciousness of kinds of qualities that are different from such well-known properties as mass, gravitational attraction, and electric charge. For example, while these commoner properties seem to be inexorably attached to the individual particles of the matter which exhibits them, consciousness seems to appear only in certain complex organizations of matter, and then only if suitable internal physical conditions are met. Although we may ultimately be able to discover the anatomic and metabolic requirements for the existence of the conscious state, this "cooperative" nature of consciousness may continue to set it apart from other properties of matter.*

It must also be recognized that "consciousness" is a word we apply to a complex of properties rather than just one. We must expect that any scheme for deriving its complete content from the physical state of the neuronal material must include relationships separately predicting a number of components. This has already been implicitly recognized as true, in references to the physical concomitants of what we imprecisely call sensations, emotions, and thoughts.

The truly unique feature of consciousness is, of course, its subjectivity. To be sure, it may be argued that all quantities dealt with by the scientist make themselves known to him, ultimately, by some sort of subjective effect. Nevertheless, for the so-called physical quantities, an important world view intervenes between sensory input and conceptual result, whereby an objective, extramental existence is ascribed to the source of the incoming impressions. Not so with consciousness. By definition it is a purely internal experience—perhaps caused by events in the external objective world but not itself a part of that world. Therefore there is only

---

* But maybe not. When subnuclear particles come together, previously nonexistent physical properties also sometimes appear.

one possible intrument for determining its existence and measuring its content: the brain—strictly, only the brain of the particular scientist who is performing the research, for he can have no way of being sure that the words spoken about consciousness by another refer to the same kind of personal awareness that *he* feels. As already observed, no individual performance or conversation would be changed in the slightest by the absence of a passive property such as consciousness is believed to be. However, because any other hypothesis would be violently inconsistent with the orderliness that so much evidence attributes to nature, few seriously question the point of view referred to earlier as the commonsense one: all human beings and other higher animals experience consciousness; it is likely that most human beings mean about the same thing by a given description of a conscious experience; it is probable that such animals as dogs, cats, and monkeys are conscious when their general appearance of alertness is similar to that of conscious human beings and that they do indeed feel afraid, angry, or happy when we think they do. With such assumptions, the behavioral responses of animal subjects as well as the reports of human patients provide the research scientist with measuring instruments suitable for use in delineating the conscious consequences of the various possible physical states of the material of the brain.

This gets us back to the nub of the argument: if the properties of consciousness can indeed be shown to be precisely determined in rigid cause-and-effect fashion by the physical state of the associated material, then conscious phenomena clearly belong in the subject matter of basic science. The unusual properties of consciousness which make it seem so different from quantities which we think we understand better do not disqualify it for inclusion. Indeed, if concepts had in the past been excluded from physics when they seemed too bizarre or hard to comprehend, there would certainly be no relativity or quantum mechanics today. And even a cursory examination of modern subnuclear physics reveals well-regarded hypotheses and theories that would seem as strange and mysterious to most of us as the ideas about consciousness that we are here considering.

In short, according to this thesis the evidence for the operation of physical cause and effect in conscious phenomena is convincing, and therefore consciousness, in the mid-twentieth century, is finally ready to make the same transition from metaphysics to physics that was set in motion for the other functions of the body in the early 1600s. Hence it is no more appropriate today to consider consciousness to lie outside the realm of subject matter suitable for investigation and understanding than it was appropriate, after the advent of Harvey, to consider the functions of the heart to be replete with mysteries that must forever lie beyond the comprehension of mortal man.

Finally, these considerations suggest that the incorporation into physics of the results of research on consciousness will not be intrinsically difficult, but will require only that the relations between specific states of matter and the resulting specific states of consciousness, as they are discovered, be included among the laws by means of which the scientist describes and predicts natural phenomena. This will constitute only another of the extensions and alterations that scientists occasionally make in their formulation of the laws of physics in order to improve its descriptive and predictive accuracy. And these additions will indeed bring about a substantial improvement, for the result will be a single body of laws, or relations among observable quantities, suitable for use in describing every aspect of human experience and behavior.

part 5

*Implications of
the Physical Explanation
of Biology*

chapter 16

The Machinelike Nature
of Man

In our study of the physical properties of organisms we learned of the strength of the modern case for a simple clear-cut conclusion: that the properties that differentiate living cells from inanimate matter result directly from the nucleic acid/protein enzyme mechanisms, and that a complete explanation of these mechanisms is possible in terms of the operation on inert chemical ingredients of the ordinary laws of physical science. We found, further, that the principles of evolution lead to a purely physical explanation of the origin as well as the chemistry of simple forms of life. Finally, we saw how the cell-diversity mechanisms have permitted the development of complex organisms, up to and including man.

The physical explanation of simple behavior was also found to lie within the capabilities of modern science. We learned that the electrochemical properties and modes of interconnection of the nerve cells account directly for the reflexes that underlie not only the autonomous regulation of physical processes that keeps higher animals alive and healthy but also much of the seemingly purposeful behavior of the lower animals. Neuronally stored response subroutines, each triggered by its own prescribed input stimuli, were also found to contribute greatly to the "lifelike" character of animal behavior.

The case for the physical basis of intelligence, while somewhat indirect, was by no means unconvincing. First it was noted that a family relationship between computers and brains was suggested by their impressive structural and functional similarities. This led to attribution of biological relevance to a network theorem proving that if there *is* a purely physical explanation of brain performance, then computerlike structures are in principle capable of precisely duplicating such performance. It was considered significant that even in their present primitive stage of evolutionary development, computers can be caused to exhibit intelligent behavior. Indeed, we finally saw there was good support for a very far-reaching conclusion: all intelligence, whether of computer or brain, is the natural consequence of the powerful symbol-manipulating capabilities of complex switching networks; therefore the ordinary laws of the physical scientist are adequate to account for all aspects of what we consider to be intelligent behavior.

This left consciousness to be explained in terms of the laws of physics. Despite its subjective nature, we found much evidence suggesting that the content of consciousness, as well as its presence or absence, is determined in detail by the physical structure and electrochemical state of the material of the brain. In the sense of reducing its properties to the direct and predictable consequences of ordinary processes, this was held to constitute a physical explanation of consciousness. It was even pointed out that there is no logical reason why the definition of physics cannot be broadened to include in its subject matter the subjective conscious experiences of higher animals.

Thus we have failed to discover any aspect of life—whether related to the origin of organisms, to their physical properties, to behavior, to intelligence, or to consciousness—whose explanation appears today to lie beyond the ultimate capabilities of physical science. In the late 1960s we seem justified in the broadest possible application of what may be called the central thesis of physical biology: that a single body of natural laws operating on a single set of material particles completely accounts for the origin and properties of living organisms as well as nonliving aggregations of matter and man-made structures. Accordingly, man is essentially no more than a complex machine.

There is, of course, nothing new about this thesis. It has been discussed since the time of the ancient Greeks. Indeed, it has long since been accepted as valid by many modern philosophers and scientists. But this book is not primarily addressed to that small and sophisticated group. And the content of most oral and written public pronouncements makes it clear that the large majority (including even some scientists and philosophers) have not yet accepted the conclusion to which our considerations have led us. Men are always reluctant to abandon any of the anthropocentric legends they traditionally employ to bolster their feelings of self-importance, and the concept of the machinelike nature of man is incompatible with a long-cherished belief in human uniqueness. Furthermore, until recently it has not been difficult, even for thoughtful persons, to conform with convention by rejecting this concept, because there has been little to suggest that it was more than one of many competing philosophic theses, all similarly unverified and seemingly unverifiable. Thus it is probably not surprising that the notion of the machinelike nature of man should still be generally rejected, as a persisting consequence of the inability of philosophic reasoning, when unsupported by corroborating scientific evidence, to overcome the powerful effects on human thought of centuries of social indoctrination.

But times have changed. What was once pure philosophic speculation has now reappeared as well-supported scientific theory. Consider this treatment, for example. Its goal has from the outset been no more than a cause-and-effect explanation of biological phenomena. The method of approach, the attitude, the language—all have been characteristic of the pragmatic scientist rather than the philosopher. We have arrived at the idea of the reducibility of biology to physics, not by abstract philosophizing, but by the most common kind of scientific reasoning—it is the most obvious and least complicated explanation that fits all the facts.

This point is important. There is more to it than simply the idea that the complete reducibility of biology to physics, with its related concept of the machinelike nature of man, is now a respectable scientific theory rather than a tenuous philosophic concept. For it is not just one of many equally likely interpretations of the results of recent scientific research; it is the *only* simple, direct, and uncomplicated interpretation of those results that anyone has been able to devise. It is true that other theories, involving one form or another of vitalistic assumption, can be force-fit into the pattern of existing knowledge by rejecting as incompletely proved the conclusions of some of the recent experiments. However, unless there is a completely unforeseen reversal of the current trend of discovery, the time is rapidly approaching when such alternative biological theories will lose the last vestiges of credibility they now retain. Thus, *with a degree of confidence not less than that which we feel in other well-established scientific theories, we can today assert that biology is indeed a branch of physical science and that man is only a complex kind of machine.*

Much could be said for terminating the discussion at this point. The biological evidence has been considered and the case made for its physical explainability. And a general conclusion has been drawn which, while obtained as a direct consequence of the pragmatic scientific approach, nevertheless is closely similar to one of the tenets of many professional philosophers. Further discussion might carry us unacceptably far away from science and into philosophy.

Nevertheless, despite the danger, it seems impossible to stop just here. For when related to conventional ideas, the concept of mechanical man has some bizarre implications. Although they are well understood by many philosophers and scientists, these implications are appreciated by few nonscientists. And they are important. Indeed, some are in sharp conflict with long-established social tradition. It would appear that anyone who has stayed with the development to this point would inevitably be interested in such related matters. Therefore, without any attempt at completeness, some of the more obvious consequences of the idea of the machinelike nature of man will be outlined in the next three chapters. It is suggested that readers to whom this kind of discussion is unfamiliar may want to defer final acceptance of what we are calling the central thesis of physical biology until they are sure that they understand the extensiveness of the resulting changes that may be required in their accustomed modes of thought.

chapter 17

Personality

To help the reader think through some of the implications of the conclusion that biology is completely reducible to physics we will employ a time-honored device—that of the *thought experiment.* This device is purely instructive, intended only to demonstrate consequences of knowledge that has been obtained and verified by other means. For such an explanatory purpose it is frequently possible to invent an idealized experiment that will sharply display some important scientific principle. Usually all that is necessary is either to ignore certain practical limitations on the capability of the real experimenter or to project the proposed test into a future period when, it is hoped, solutions will have been found for these practical problems. Thus, in Archimedes' description of his famous thought experiment on levers, "Give me a place to stand on and I can move the world," he expressed in vivid fashion a valid consequence of his theory of static forces even though he described an experiment that could never actually be carried out. And Galileo was able to predict what would happen (if his theories were correct) to objects falling in a vacuum, although it would be many years before technology would permit actual performance of the tests he described. In the present case, we shall not hesitate to ascribe to future surgeons and biologists capabilities far beyond those of today, if important aspects of our scientific concepts can thereby be illustrated.

Let us now consider several thought experiments that display some of the more startling implications of our theory, bearing on the meaning of such concepts as "individual" and "personality."

Man-made Men

Machines are made by men. If man is only a complicated kind of machine, should it not be possible, at least in principle, for men to manufacture other men—not just in nature's way, but by synthesis from completely inert, inorganic ingredients?

Modern genetic research provides a rather clear line of speculation as to the way in which human machines might one day be synthesized. The key to the success of the manufacturing process will consist of a method of constructing the DNA molecules, which, as we found in Part 1, contain in their complex architecture an atomically coded book of instructions for the anatomy and metabolism of the organism. We have learned that simple forms of DNA can now be synthesized out of the inorganic ingredients of the chemist's stockroom and that progress is also being made in analyzing the DNA molecules of living organisms and in devising techniques of duplicating their complex structure. Because of such developments, many scientists are convinced that synthesis of the complete nucleic acid book of instructions will someday be possible.

Of course, genetic synthesis is not the only problem that must be solved. DNA molecules alone, for all their importance, cannot develop into a complete organism. Probably the most practical approach for the designer of our future man-manufacturing plant to take will be to incorporate the DNA into the synthesized equivalent of fertilized egg cells and then to nourish the egg cells in an artificially produced environment similar to that now provided in the mother's uterus. Although no one would contend that the manufacture from inorganic ingredients of the non-nuclear part of the ovum or the duplication of the uterine environment will be easy, there would probably be general agreement that these accomplishments should lie within the capabilities of any scientists who have been able to synthesize the DNA content of the human cell. In principle, at least, it is no harder to visualize the one kind of achievement than the other.

We come, finally, to the point of these imaginings: the creatures manufactured as described will be perfectly normal individuals—completely indistinguishable from human beings produced in the usual manner. The fact that they originate out of the inorganic ingredients found in the chemist's laboratory and develop without the direct intervention of any living organism is entirely irrelevant to the final result: if the anatomy and metabolism of these manufactured structures are identical with those of ordinary human beings, they will *be* ordinary human beings. They will be physically human; their behavior and intelligence will be human; their properties of consciousness will also be human. The central thesis of physical biology permits no other conclusion than this.

Conscious Computers

In their relaxed moments, computer scientists are prone to speculate on the great advances their technology could make if only it were possible to incorporate into their equipment some of nature's computer components. They know, for example, that the memory capacity of 1 cubic inch of the neuronal material of the brain of a monkey or human being exceeds that of a roomful of electronic components. They are therefore intrigued by the possibility of someday devising methods of coupling their electronic devices to the flesh-and-blood organs of higher animals. With better understanding of how nature's computers work, there seems no fundamental reason why the problems associated with such interconnection cannot ultimately be solved. And the success of such experiments as those pioneered at Western Reserve University in Cleveland, in which the brains of monkeys have been removed from their bodies and kept alive and functioning for twenty-four hours or more, suggests that the physiological problems involved are also susceptible of solution.

Thus, in accordance with the general rules of thought experimentation, it is reasonable for us to assume that such intermingling of electronic and organic components will eventually be accomplished and to speculate on the consequences. Let us suppose, for example, that the brain of a monkey or human being is incorporated as a working part of an advanced type of computer. Will it be conscious? We must conclude that it will be if the environmental conditions are made similar to those which accompany consciousness in the normal brain. This will of course be equally true if the organ employed does not come from a complete organism, but is instead grown in a parts-manfacturing plant, so to speak, with the generating DNA and egg-cell material specialized to result in the development of brains only.

But if animal brains can form useful components of future computers, it is likely that more specialized structures will be even better suited to the specific requirements of the associated electronic equipment. Perhaps, for example, only the memory portions of the brain will be needed by the engineers. Scientists who have succeeded in synthesizing entire human beings should have no great difficulty in modifying their technique to produce such specialized components. Will such humanoid but not human structures also be conscious? The answer must be "Perhaps." It all depends on whether the organization of matter and the associated electrochemical activity achieved in the synthetic components meet the conditions that future research will show must result in consciousness. Until these new laws of physical science are determined, we will have no way of predicting how much the structure and metabolism of the brains of higher animals can be changed without the loss of the capability for consciousness; however, we must assume that some changes are permissible, perhaps very extensive ones. Indeed, our thesis requires that we keep an open mind as to the possibility that among the wires and transistors of existing electronic computers, there already flickers the dim glimmering of the same kind of personal awareness as that which has become, for man, his most precious possession.

Surgical Creation and Fusion of Personalities

If a discussion of the use of animal brains as computer components qualifies as imaginative speculation rather than sheer fantasy, it is only because of advances already made in the research laboratory. The Western Reserve successes have been cited as a source of confidence in the ultimate tractability of the surgical and physiological problems involved in keeping isolated brains alive indefinitely. Similarly, growth in understanding of the electrical nature of the nerve impulse has encouraged us to believe that one day it may be possible to send meaningful information into the sensory nerves and extract useful signals from the motor nerves of a detached brain. But these anticipated future developments are also all that stand in the way of realization of a favorite theme of science-fiction writers—that of the dead genius whose brain is kept alive to continue to exercise its influence (usually evil) on the surviving cast of characters. We shall find something like this science-fiction theme a convenient point of departure for thought experiments illustrating additional consequences of the thesis of the machinelike nature of man.

Specifically, let us imagine experiments on a human brain that has been surgically removed from its body and provided with a combination of chemical fluids and electrical inputs which keep it alive and conscious. As to the question, "Conscious of what?"

we will assume that the brain will, in the first place, have its past memories to use as subject matter for thoughts. In addition, if the electronics engineers on the project have done their jobs well, we may assume that televisionlike signals applied to the optic nerve will supply the brain with the equivalent of vision, and that an electronic lip/tongue/throat mechanism operated by impulses coming from the motor nerves will permit it to give vocal expression to its thoughts.

Perhaps a word of reassurance is in order regarding the seemingly gruesome aspects of the postulated experimental arrangement. We do not necessarily have to feel sorry for the disembodied star of our planned production. In the first place, we will certainly control the conditions so that it feels no physical pain or discomfort. Furthermore, from the work on pleasure and punishment centers, we know that we can also control its emotional state, making it feel continually relaxed, happy, or even ecstatic simply by arranging for suitable patterns of electric current in selected regions of the brainstem. Indeed, if such experiments ever really become possible, a major problem may be the selection of lucky winners from the many who volunteer for disembodiment because of their wish to achieve a happier state of existence than that available to them by ordinary means.

The first significant experiment we can perform with our new apparatus consists simply of confirming that the arrangement works as required by our thesis and as just described—that the isolated brain does continue to function and give evidence of conscious thought. But other interesting experiments are also possible. Let us consider, for example, an application of the split-brain technique of R. W. Sperry's Caltech group to our new experimental arrangement.

It will be convenient this time to have a name for our subject: let's call him Tom. After his isolated brain has been successfully and happily established in its new laboratory environment, let us, using general anesthesia, surgically separate the two halves of the brain and establish each in its own separate arrangement of supporting equipment.* From the split-brain observations that have already been made we know what to expect when the anesthesia wears off. We will then have two Toms instead of one: the two half-brains will have exactly the same memories and each will feel itself to be the original individual. Here will be the first case of twins really entitled to the adjective "identical." And clearly we will be able to say correctly that, in this instance, a new personality has been created by surgery.

Now let us reverse the surgical process by putting the two brain halves back together (a substantial chore, involving the reconnection of more than three hundred million nerve fibers). The result of the restoration will, of course, be a single Tom. By surgery, two previously separate personalities will have been blended into one.

The relevant point could be made even more dramatically by an experiment involving two subjects, Tom and Dick. After splitting both brains, the left half of Tom's organ could be combined with the right half of Dick's. To be sure, in this instance the combination might not work. The differences in the patterns of thought of the two individuals might result, upon interconnection, in electric currents and chemical reactions preventing any kind of coherent performance of the hybrid organ; the combination might even be fatal to both half-brains. But if the combination *did* work, our theory would predict an interesting new kind of blended individual, possessing in one single consciousness the memories, learned habits, and senses of identity of both Tom and Dick. This would be schizophrenia with a vengeance!

* We may ignore certain practical physiological problems that have so far limited the amount of brain splitting to the cortex and the top parts of the brainstem and assume future surgical development permitting healthy functioning of completely separated half-brains.

Made-to-order Personalities

If we can develop our technology to a point permitting the injection of meaningful environmental data into the isolated brain (we have used the example of visual input) we should be able to go further and record complete new patterns of memory in the subject. By such a method the isolated brain could be provided with detailed recollections involving people who never lived and events that never occurred. If properly implanted, these recollections would be as real to the subject as those which had actually taken place during his period of corporeal existence. We can even imagine the creation of made-to-order personalities, involving the implantation of a complete complement of memories, habits, and thought patterns in the virgin material of a factory-made brain.

Indeed, if we give free rein to imagination (and we may already be centuries ahead of the state of the art), we can even conceive of a sort of copying of a real personality. The pertinent analogy is an electronic computer which is caused to disgorge the complete contents of its memory store for use in setting up a second computer to be functionally equivalent to the first. Some unusually interesting consequences of our theory of the machinelike nature of human beings emerge from thought experiments in which a similar technique of recording and reproducing the contents of the memory store is assumed applicable to the brain.

Specifically, let us imagine performing an experiment on a normal adult human being comprised of the following steps: (1) Employing suitable anesthesia, we connect appropriate apparatus to the subject's brain and make a permanent record of the complete contents of his memory (including physical and mental habits, of course). (2) We transfer this record to the neuronal material of a fresh, factory-made human brain. (3) We replace the natural brain of the subject by the newly processed factory-made article, making suitable connections to all incoming and outgoing nerves. (4) We allow the subject to awaken. The pertinent question is "What, or rather who, is the result of this series of steps?"

To all outward appearances the individual is unchanged by the operation. His memories, habits, patterns of thought, skills, worries, fears, and satisfactions are exactly as they were before. These are all determined by the structure and chemistry of the material of the brain, all essential aspects of which have been accurately established in the synthetic organ by the postulated duplicating process. To be sure, the material of the brain is all new, and our conventional modes of thought almost automatically cause us to conclude from this that the old individual is gone. But is he, really? After all, metabolic processes continually replace the molecules in the normal brain; hardly any material has been with any of us for more than a few weeks. Yet we do not believe that this causes us to become new individuals. Logic, as well as the thesis of the machinelike nature of man, would seem to require us to take a similar point of view with respect to the experiment we are considering—to conclude that the personality of the subject not only *appears* to be unchanged but *is* unchanged, despite the complete replacement of the material of his brain.

There is an amusing "practical" application to future space travel of the imaginary operative technique just described. Space scientists, who also sometimes speculate far into the future, generally conclude that it may require thousands of years for vehicles to reach planets containing other civilized beings, even if speeds approaching the velocity of light are attained. The problem posed by the relatively short life span of the human astronaut is an obvious one. The solution, according to our line of speculation, could be as follows: Record the contents of the astronaut's brain and his DNA genetic specifications on durable, long-lived storage material. Arrange a mechanism in the spaceship which, at the future time of interest, will automatically combine suitable inert chemicals and subject them, under control of the recorded genetic specifications, to the physical manipulations involved in the man factory described earlier. Then, by an equally automatic process, impress on the virgin material of the new creature's brain the total recorded contents of the brain of the ancient astronaut. Finally, allow the finished product to awaken in time to meet the inhabitants of the planet to be visited. Pure science fiction though it may be, this imaginary sequence of events illustrates again one of the most bizarre consequences of the idea of the machinelike nature of man. For even though thousands of years should intervene between the recording of the design details of the astronaut and his subsequent duplication, we must conclude that it would be as accurate to call the two organisms the same individual as it is to call any one of us today the same person he was yesterday.

The Transience of Consciousness

Other conceivably possible future experiments can easily be invented to demonstrate implications inherent in the concept of the machinelike nature of man. This will be left as an exercise for the reader, since the illustrations already presented have brought out the important points. It will be best for us now to devote our attention to elaboration of the meaning of some of the material we have already considered. For the conclusions related to consciousness are so different from conventional ideas as certainly to have raised some perplexing questions in the minds of many readers. For example, when Tom's brain was split to make two individuals, mustn't one of the half-brains have been the *real* Tom, and the other only some kind of imitation? And in the last example considered, where was the astronaut *himself* and what was his condition during the thousands of years that separated his two incarnations? Was he dead? If so, did he then come alive again? Or was he alive all the time, in some mysterious kind of dormant state?

According to our thesis such questions, while subjectively compelling, are objectively meaningless. They arise from an erroneous assumption that underlies traditional ideas. This assumption so permeates the meanings we attach to personal pronouns like "I," "you," and "he" and related terms like "person" and "individual" and "alive" and "dead" as to embed self-contradictions in the very language we must use to discuss experiments such as those described in this chapter. Stated another way, confusion can easily result when we use words that by definition attribute an independent existence to consciousness in discussing a theory in which consciousness is only a transient property of the current state of organization and electrochemical activity of matter.

In terms of the view of consciousness as a derived property of matter, such questions as those pertaining to the life or death of our astronaut during his long space journey belong in exactly the same category as the old riddle, "Where does a light go when it goes out?" The light is simply a phenomenon which automatically results from the right kind of electrical activity in the right kind of filament material. It has no independent existence of its own. When the physical conditions are suitable, the light is "alive"; when not, it is "dead." But we do not think of the light we see on a second turning on of the lamp as being the same light we saw earlier and as having meanwhile been somehow in existence though invisible. Similarly with the ancient astronaut: in his original incarnation, at least during his waking periods, the organization and physical state of the material of his brain caused the phenomenon of consciousness to exist; when, much later, the automatic factory in the space vehicle assembled material substance in accordance with the detailed stored specifications of the astronaut, the state of consciousness again appeared, with a content completely determined by the stored record—that is, with a complement of memories, habits, and thought patterns, or personality, identical with that of the human being who had existed so long ago; in the intervening thousands of years, the inert recording of the pertinent specifications of the astronaut's body and brain did not meet the physical conditions for the appearance of consciousness—"the light was out." There is really no more to be said. Life and death are seen, on this level of discussion, to have significance entirely different from that commonly attributed to them.

Indeed, in terms of our traditional emphasis on the consciousness of the individual, it might be said that we are as dead as we will ever be each time we go to sleep or are anesthetized. The point can be made even more dramatically: since the consciousness-producing electrochemical processes in the material of the brain decay in a few seconds if not renewed by suitable external stimulation, it is justifiable to assert that the individual, or personality, or soul, we attribute to consciousness has no greater longevity than this. In such terms, each of us dies and is reincarnated many times each minute.

Unless very carefully considered, such statements must appear either incredible or foolish. They *feel* wrong. Probably our most pervading mental impression is that of unchanging personal identity. Each of us *knows* himself to be the same person he was yesterday; his memories, habits, thought patterns, and feelings have a sense of continuity and permanence that seems to belie completely the kind of transience of conscious identity being described here. But this feeling, like other subjective impressions we have earlier found to be unreliable, does not form a sound basis for a scientific theory. For there is no objective reason to ascribe anything except the present tense to the content of consciousness. To be sure, at any instant one's thoughts may include components related to past events and previous intellectual activity, but he may be aware of these previous matters only because, *at this instant,* his neuronal activity is reproducing them. If we could by surgical means make suitable instantaneous modifications in the patterns of synaptic interconnection of the neurons, there is no reason to doubt that we could equally instantaneously change the personality of Tom to that of Dick. And there would be absolutely no way for the new Dick to realize that, a second ago, he was Tom: after the operation he would be as convinced that he had always been Dick as he was previously convinced that he had always been Tom. This would be inevitable, if indeed the content of consciousness is solely determined by the current physical structure and state of activity of the brain. In the personality of the normal individual there *is* continuity, but this would be a natural consequence of the normal absence of abrupt physical change in the organization of the brain. It need have nothing to do with any quality of permanence in the property of consciousness itself.

The Major Conclusion

With this discussion of the transience of consciousness we have explored the most unconventional of the philosophic consequences of our central thesis. If the reader can accept the conclusions of the last few paragraphs, he should have no difficulty with the other implications of the theory. But the conclusions about consciousness are an essential part of the argument; they cannot be ignored as irrelevant to the other major consequences of a thesis proclaiming the complete reducibility of biology to physics. For, as we have seen, such a thesis requires that the properties of the conscious state be completely determined by the physical condition of the related material particles. Consciousness is thus denied the independence of matter that is traditionally assumed; the consequences we have just discussed follow inexorably.

Again it should be recalled that we have examined considerable experimental evidence directly attesting to the physical control of the existence and content of consciousness. Such evidence constitutes an important part of the case for the central thesis of physical biology. Those who find the argument for the thesis convincing cannot escape the resulting consequences. Strange though some of them appear to be, they are the price we must pay for a world view in which all human experience is basically lawful and orderly.

chapter 18

Free Will

In the context of a completely physical biology, free will poses no problem—it simply doesn't exist. Obviously, it cannot, if conscious personality is no more than a derived, passive property of certain states of organization and electrochemical activity of the neurons. On this basis our thoughts and actions must be as rigidly controlled by the operation of inexorable physical law among the material particles of the universe as is the movement of wind and wave.

Logically, it is not clear that anything more need be said about free will in a treatment such as this one, which is concerned with philosophic concepts only when they are conspicuously exposed by the trend of scientific discovery. However, this subject has been discussed for so long by so many that some elaboration may be indicated in order to prevent confusion with other and better-known points of view. Three topics that frequently arise in modern discusions of free will are briefly touched upon in the following three sections.

The Principle of
Indeterminacy

In traditional debate, the terms "free will" and "determinism" have generally been considered to refer to accurately antithetical and mutually exclusive philosophic concepts. Thus any point of view denying the existence of free will, such as the thesis we are discussing, is usually referred to as a deterministic point of view. But there is another common definition of determinism: the concept that if at any instant the physical state of every particle of matter in the universe could be precisely described, then every slightest detail of future history would in principle be *predictable,* including the actions, words, and thoughts of every individual. Because the two definitions do not necessarily mean the same thing, confusion has sometimes resulted in discussions related to free will and physical biology.

It is, of course, the quantum-mechanical principle of indeterminacy that has clouded what used to be considered a clear-cut issue. The twentieth-century formulation of the laws of physical science possesses a statistical character limiting the precision with which predictions can be made. And many scientists believe that this results, not just from an incomplete understanding of the natural laws that may one day be set straight by further discovery, but from a degree of fundamental unpredictability in nature itself.

Thus the concept of biology as a branch of physics is not necessarily deterministic in the sense of implying the completely detailed predictability of the future behavior of organisms. However, it is still deterministic in the sense of providing no way for such a nonphysical agent as free will to participate in the control of thought and behavior.

Indeed, there is only one way that the principle of indeterminacy could have any bearing on the case supporting the reducibility of biology to physics—by extensively weakening the cause-and-effect relationships provided by physical science. After all, the argument for the completely physical nature of biology rests on the ability of physics to provide convincing explanations of the phenomena of life. Any vagueness or indeterminacy in these explanations is bound to reduce the conviction they carry.

Fortunately, quantitative considerations easily dispel this kind of concern. For the indeterminacy of the laws of modern physics is of a highly restricted nature. It appears to a detectable extent only in sub-submicroscopic situations, and even then its limits are rather precisely prescribed. Indeed, the quantum-mechanical weakening of cause-and-effect relationship is negligible in all of the evidence we have considered for the adequacy of physical explanations of biological phenomena. The number of material particles involved in biologically significant structures is so great and the resulting uncertainty of physical prediction is so small as to render it improbable that this kind of consideration will ever be important in the science of life.

Predictability and Compulsion

In recent years some philosophers have chosen to redefine free will so as to make it compatible with a purely physical concept of biology. They point out that when any of us decides whether or not to go to a ball game, immediate outside influences alone do not predetermine our decision. Instead, our own unique system of stored recollections, habits, and values enters into the determination. To be sure, the resulting decision is in principle predictable (ignoring the slight effect of quantum-mechanical indeterminacy), because it depends rigorously on the state in which our billions of neurons have been left by past experience (and of course on the present patterns of neurological stimuli). But there is no real compulsion, in the sense of a specific decision imposed by some outside agency without regard for the prejudices and desires of the individual. In this sense, free will is said to operate.

There can be little logical objection to any definition. However, this use of the term "free will" seems to imply some kind of basic difference between the physical processes in the material particles making up the body of the individual and those in the material particles lying outside the body. To one who really accepts the physical basis of biology the kind of compulsion exerted by the bars of a cage and that exerted by the established neuronal patterns of an agoraphobiac * seem to have much in common. One involves the workings of the ordinary laws of physics in the material of the cell bars, the other their workings in the material of the brain. Both result in keeping the victim confined. And the amount of free will of the kind describing the possibility of an action not completely dictated by the immediate configuration and activity of material particles is the same in both cases—none at all.

The Subjective Evidence for Free Will

There is no doubt that free will *seems* like a very real thing to most of us. This can be demonstrated in many ways. If, for example, the suggestion should be made here that the reader look up, close his eyes, and deliberately shift his conscious attention to another topic of his own choosing, he would probably describe the result about as follows: *"I considered* the suggestion; *I decided* to accept it; *I directed* my muscles to lift my head and close my eyes; *I reviewed* various items from my stored memories; *I chose* to think about" such and such a topic. The personal pronoun "I" and the verbs "considered," "decided," "directed," "reviewed," and "chose" paint a clear picture of an independent executive agency that is freely considering alternatives, making decisions, and issuing commands. As observed earlier, the normal terminology of language presupposes free will. Every time we formulate such a statement as the example given, we uncon-

---

* One with an abnormal fear of open spaces.

sciously reinforce the conviction we already have in the potency of the nonphysical agent "I."

But, at the expense of some linguistic clumsiness, the same sequence of thought processes could also be described deterministically, without introducing the slightest incompatibility with the nature of the subjective experience. The first part of the description would be easy, involving only replacing "I" by some such phrase as "the switching circuit in the brain." For Part 3 is considered, in this discussion, to have established that the purely physical properties of the neuronal structures are adequate, without further stimulus than that provided by the outside environment, to result automatically in such intellectual actions as considering, deciding, directing, reviewing, and choosing.

The remainder of the deterministic description would have to deal with the conscious concomitants of what we are now thinking of as automatic neuronal events. For example, to the statement, "The switching circuit in the brain considered the suggestion," we would need to add something about as follows: "A few of the many neurons involved in this process happened to lie in a region of the brain where the combination of structure and electrochemical activity resulted in conscious effects. Those wired into the circuit so as always to fire when the brain mechanisms compare the advantages and disadvantages of alternative courses of action produced a familiar feeling known from experience to be related to 'considering' or 'deciding.' Others resulted in conscious components related to the details of the specific suggestion being considered."

In similar manner, the other statements could be expanded to provide for the conscious aspects of the intellectual processes we are considering. In each case, reconciliation of the deterministic view with the feel of the conscious experience would involve the conclusion that we ordinarily make a certain kind of mistake in interpreting that experience. We attribute the familiar and identifiable conscious feelings associated with considering, deciding, directing, reviewing, and choosing to the operation of a nonphysical executive agent. But all that any such feeling needs to mean is that the automatic mechanisms of the brain are engaged in the related process. There is absolutely nothing in the conscious experience that is fundamentally any less compatible with this interpretation than with the time-honored concept that the physical brain is being controlled by a nonphysical "I." To be sure, the deterministic concept does not *seem* right to us. But this is only because of long-practiced contrary ways of interpreting conscious phenomena. As children, we were thoroughly indoctrinated with the conventional point of view that certain kinds of subjective sensations indicate the exercise of nonphysical control over our thought processes; since then we have reinforced this prejudice countless thousands of times by noting its consistency with detailed subjective experience. It is doubtful whether there exists today a single adult who has been reared from infancy under the influence of the deterministic view of consciousness and who has therefore reinforced *this* initial prejudice by countless thousands of observations of *its* consistency with subjective experience.

chapter 19

Social Attitudes

The concept of mechanical man must ultimately give rise to important changes in social attitudes. Of course, the changes will result from the *idea* that man is a machine, rather than from the *fact* of his machinelike nature. For our long ignorance of what we really are has subtly influenced the ways that the laws of physics have exercised their inevitable control over us. The point is that, although a machine, man is a sensitive and flexible machine whose behavior is influenced by the concept of his own nature that is stored in the switching network of his brain. If the trend of scientific discovery we have discussed continues until the machine-like nature of man finally comes to be widely accepted, there could ensue important changes in man's attitudes and institutions. It is these changes we shall now consider.

Religion

A state of consciousness whose existence and properties are completely determined by the operation of the laws of physics in the material of the brain, and which experiences death and re-birth several times a minute, would appear to leave conventional religion very little to work on. With no independently surviving soul to reward or punish in an afterlife, there can be no purely religious compulsion toward acceptable social behavior. Indeed, the strict physical determinism underlying the machinelike nature of man will render impossible traditional judgments as to who should be rewarded and who punished, even if there should be a suitable object on which to carry out the sentence. For the typical religious concepts of Right and Wrong are hardly reconcilable with a situation in which every detail of the behavior of each individual is determined by the physical facts of his heredity and environment.

In addition to giving up its traditional weapon for compelling moral behavior, religion will have to make drastic modifications in its theology. There is obviously no room for a personal God in a world that is rigidly obedient to inexorable physical laws. Miraculous suspension or modification of such laws in behalf of a group or individual, perhaps as the result of a direct prayerful appeal to some higher power, will be even less credible than it is now. Equally unacceptable will be the prophets and other super-human individuals who typically strengthen the popular appeal of religious dogma by surrounding it with an attractive aura of unworldly mysticism.

This is not to say that complete atheism will be required. We have already observed that while science seems well on its way to the elimination of basic unlawfulness from human experience, it cannot eliminate the underlying mystery. There will be no reason why the term "God" cannot still be used to denote the seemingly inexplicable origin of the laws and particles of physics.

Morality of the Individual

With the general recognition of the meaninglessness of the traditional and essentially mystical concepts of Right and Wrong, will immorality and crime run so rampant that human society itself will become impossible? Almost certainly not. The concept of absolute, God-given standards of morality is by no means essential to a lawful and orderly society. To be sure, the religion of a country, whether it is Christian, Buddhist, or Mohammedan, is frequently given credit for the generally moral behavior of its people. However, it is exceedingly doubtful whether this credit is deserved. Indeed, there is no evidence that the incidence of crime and immoral behavior is any higher in the several large atheistic countries that exist today than in the Christian democracies.

According to our mechanistic point of view, a tendency toward moral behavior is a genetically determined, evolutionarily developed physical property of the human animal, just like the number of fingers and the size of the brain. For man's dominance over other species is in large part the result of his social attributes—only by living and working together is it possible for men to secure the advantages flowing from the knowledge obtained by others and for successive generations to progress by accumulation of such knowledge. But living and working together is impossible unless each individual is willing to accept some restraints on his own activities for the good of society, that is, to exhibit moral behavior. Thus man's superior accomplishments became possible only when the trial-and-error processes of evolution finally succeeded in providing him with a built-in scheme of nervous-system design that gave his behavior an adequately social bias.

We do not yet understand well enough the complicated machinery of the brain to be able to determine just what kind of interconnection among the neurons is involved in making man a social animal. However, introspection as well as observation of human nature suggests that the underlying principle is a mechanism that causes each to derive pleasure from the indications of approval he receives from other human beings and pain from their expressions of disapproval. Probably no more than this is needed to cause us to aggregate in tribes, cities, and nations and to motivate us to subordinate some of our own personal desires and needs to those of the community.

Of course, men do occasionally lie, steal, commit murder, and perform other antisocial acts. However, this is not inconsistent with the presence in the genetic make-up of a general *tendency* toward socially approved behavior. We found earlier that even the nervous system of the lowly earthworm possesses several different patterns of response to the environment, the resultant of which sometimes impels it in one direction, sometimes in another. It is probably not surprising that the much larger repertoire of human response patterns can produce occasional actions ranging in social value all the way from that of treason to that of patriotic self-sacrifice.

While the genetic basis of individual morality may prevent any catastrophic social consequences from arising out of the elimination of religious compulsion, this is not the same as saying that there will be no adverse consequences at all. It is probable that many men are now urged into more socially acceptable behavior than they would otherwise exhibit by concern for the future well-being of their "immortal soul." And even those who are not impressed by such mysticism may yet be caused to give their innate moral tendency a more prominent role in their personal decision making by the social exhortations, in part religiously inspired, that are so liberally scattered through what they read and hear. Ultimate recognition of the machinelike nature of man will certainly discredit the religious imperative that now helps hold the first class of people in line, and it may also weaken the institution of the church enough to eliminate most of its exhortative influence on the second. But there is no reason to expect society to allow such weaknesses to remain uncompensated. For recent world history suggests that governments will not find it beyond their capability to fill the one need by investing their own institutions with a semireligious aura that the unanalytic can worship, and to fill the other by intensification of propaganda emphasizing the popular approval to be expected from behavior acceptable to the state. Whether this will make the citizen's life more or less pleasant is not our concern here. At issue is only the question of the continued moral behavior of individuals after they have learned that men are machines. The point is not that traditional religion currently plays no part in preserving the fabric of society. It is, rather, that the drastic modification or elimination of religious concepts by the general acceptance of the machinelike nature of man is not likely to lead to anything like an explosion of crime and immorality. Our strong innate compulsion toward moral behavior combined with the flexibility available to social institutions can confidently be expected to prevent such a result.

Individual Ambition and Accomplishment

It can be argued that discussion of the probable morality of future behavior is pointless because general acceptance of the machinelike nature of man will result in there being little purposeful human behavior to make moral judgments about. After all, if we really believe that each of our acts and thoughts is completely determined by outside physical events beyond our control, won't we simply sit down, relax, and wait for nature to have its way with us?

Fortunately, there is evidence that this won't happen. For there are many who are already convinced of the validity of determinism. In particular, a large proportion of practicing scientists hold this view. Yet few would suggest that such people are more lazy or contribute less to society than those who believe in the efficacy of traditional free will.

Indeed, a little thought will show that a belief in determinism could not greatly influence most of our actions. To illustrate the point with an extreme example, we know that a starving man would hardly be deterred by his philosophic convictions from going in search of food. More generally, the complex drives and compulsions which operate to determine our acts and decisions must always be with us. It is possible that they can be modified somewhat, but it would be remarkable indeed if *any* idea could greatly diminish the innate energy and ambition characteristic of our species. This would imply a serious defect in the human nervous system, which would probably have led to man's extinction long before now, in the tough evolutionary competition for survival.

And our conviction that a belief in determinism will not destroy the ambition of most individuals need not rest only on evolutionary argument. For such a belief is essentially neutral: it should not bias the thought mechanisms one way or the other in the vast majority of the decisions that have to be made. Since we cannot throw up our hands and forsake our normal ambitions without making a positive decision to do so, there is no way that a belief in determinism, per se, can lead to this result.

Of course, this is not quite the whole story. Even though appreciation of his machinelike nature might not generally cause an individual to lose his drive and ambition, it could still constitute an excuse for taking the easy way out in a situation requiring difficult or unpleasant activity now to ensure social approval or other long-term reward later. In fact, observation of normal human performance leaves little doubt that determinism *will* be used in this way, to rationalize decisions that we want to make. But we are already so good at finding excuses to justify our illogical decisions that the addition of one more to our repertoire should not make much difference.

In short, some decrease in ambition and productivity may result from the general acceptance of the machinelike nature of man, but probably not much.

Treatment of the Individual by Society

We have devoted some attention to the probable effects of the new ideas on the behavior of the individual toward society. It is also pertinent to inquire as to likely changes in the attitude of society toward the individual.

Of course, laws and rules governing its members will still be established and enforced by the community. And since the survival of organized society requires it, something like the golden rule will continue to be put forth as the ideal standard of individual conduct, just as has always been true in successful human societies, even those formed by isolated primitive tribes. These basics cannot change. What can change is the attitude of society toward those who break its laws.

The religious component of the theory of punishment will of course be particularly vulnerable. The idea that it is somehow Right to punish anyone who violates "God's law" will obviously not survive general acceptance of the machinelike nature of man. The final disappearance of this vengeful concept should remove a confusing element that frequently causes society to react less than intelligently to the transgressions of its members. To be sure, those who cheat at cards may still be snubbed by their neighbors, and murderers may still be put in jail. The practical value to organized society of punishment, to deter others as well as the one punished from future antisocial behavior, will prevent its abolition. However, the degree of social ostracism, in the first example, and the length of incarceration and nature of the jail experience, in the second, will hopefully be little influenced by the feeling that the transgressor "is bad" and therefore "deserves to be punished."

Although the general acceptance of the ideas under discussion will undoubtedly lead to substantial changes in men's treatment of other men, the changes will be markedly less than they would have been some years ago. For the so-called humane movement related to the punishment of lawbreakers, which has been under way for a century, leads to precisely the same consequences as the "mechanical" movement we are now contemplating. Indeed, despite the incompatibility of their labels, both movements have the same basic philosophic premise: that man is the product of his heredity and environment. The corollary is that the transgressor should be subjected only to whatever form and degree of punishment best contributes to the good of society. The humanitarian would say that it is wrong to be more punitive than this; the mechanist would call it pointless. The end results are the same.

Objectivity of Human Thought

An important feature of human thought is a tendency toward the classification of concepts and objects as right or wrong, good or bad. On hearing an opinion stated, we must immediately decide whether we agree or disagree. We are rarely neutral about television programs, rock-and-roll music, civil rights demonstrations, or Charles de Gaulle. We are for capitalism and against communism, or vice versa. Few of us can read the newspapers very long without forming a mental table of foreign governments we like and others we dislike. And our own government's course in Vietnam is either the skillful response of wise and humane leaders to a difficult situation or the reckless arrogance of a power-mad administration.

It is possible that our tendency toward black-and-white categorization is related to the ease with which any switching network, like that comprised by the neuronal material of the brain, can shunt incoming signals into separate and discrete storage positions. Whether or not this is a valid conjecture, it seems clear that the ability to make such classifications must have been useful to primitive man. It was necessary for him to identify berries as edible or poisonous, animals as innocuous or dangerous, human beings as friends or enemies. Survival itself depended on the accuracy, and sometimes on the rapidity, of such judgments. The normal processes of evolutionary selection would inevitably have led to a nervous system with a well-developed capacity for dichotomous classification.

Unfortunately, the decisions modern man is called upon to make are frequently more difficult than determination of the edibility of a berry or the aggressiveness of an animal. More and more our safety and well-being depend on correct judgments about complex social situations involving other individuals, groups, or nations. In making such judgments too strong a tendency toward dichotomous classification may well have negative rather than positive survival value.

Consider, for example, how Americans interpret news about Red China (or how the Red Chinese must interpret news about America). For reasons related to our view of recent world history, most of us decided quite early that we didn't like the new Chinese regime or believe there was much that was good about it. Once this decision was made, even tentatively, a dichotomous-classification mechanism in our brains was set to operate: from then on we had at least a small tendency to assign to the unbelieved and rejected category anything we read or heard that favored Red China, and to the believed and accepted category anything unfavorable. This bias was automatic and largely unconscious. We may have thought we were interpreting the news accurately; but to the extent that our suspicion and distrust were already aroused, we could not do so. As a result, we distilled from the news additional "evidence" for our already negative convictions. Because of the further strengthening of our classification mechanism caused by this selected "evidence," the next reports about the Chinese Communists were even easier for us to interpret, and all difficulty in separating "truth" from "fiction" may have vanished with the arrival of still later items. For the effect of such a process of mentally editing the news and reinforcing established convictions with the credible residue is strongly cumulative. The growing body of accepted "evidence" of the perfidy of our enemy makes it progressively harder for us to believe anything favorable about him. A point can finally be reached at which we will, in all sincerity, attribute the vilest of motives to his most innocent words or deeds.

Obviously, similar mental processes can color the attitude of a liberal toward conservatives, a labor leader toward management, a Christian toward heathens, or an adult toward teenagers. Unreasonably favorable attitudes can also be formed, when the initial bias is toward the rejection of the unfavorable and the acceptance of favorable news about a thing or concept. The superiority of our country, of our religion, of our children—all can be overestimated as a result of the ease with which the automatic mechanisms of the brain put the stamp "credible" or "incredible" on incoming information in obedience to long-established, cumulatively reinforced, but essentially inaccurate sorting rules.

It is clear that this phenomenon must lower the average logical quality of our opinions. Today this can be a serious matter, for preservation of peace and prosperity in the modern world of conflict, at best a touch-and-go proposition, is made more difficult by any lack of logic in our conclusions and decisions. Society would obviously be benefited by anything that could improve the average quality of the judgments of its members. In particular, there would be merit in changes in thought patterns making us less inclined toward the quick and insufficiently analytic formation of overall value judgments.

Any hope for improvement must rest on the fact that our habits of thought are influenced by environment as well as by heredity, so that we may have some capacity to modify our innate tendency toward black-and-white classification. Most hopeful of all would be the discovery among existing social attitudes of an influence tending to aggravate the problem, which might conceivably be weakened or eliminated. This gets us back to the point of this discussion. For there *is* such an aggravating influence, and it should be weakened considerably by general acceptance of the machinelike nature of man.

It is the disappearance of the mystical concept of Right and Wrong which may result in significant increase in the logical content of human thought. Despite the enlightenment of our age compared with that of previous periods, this concept still has great influence on our thought patterns. As only one example, nearly everyone is taught to obey the Ten Commandments (or the equivalent), not just because failure to do so will cause society to make his life miserable, but because in some religious or mystical sense it is Right to obey them, Wrong to disobey them. In fact, much of what we read and hear is permeated by an emphasis on absolute moral judgments of one form or another. Inevitably this must react back on our sensitive nervous systems to influence our habits of thought. It increases our own use of unanalyzed moral judgments, of course. In addition, it must tend to reinforce all the essentially similar patterns of neuronal inter-connection in the brain underlying the neat assignment of objects and concepts to absolutely approved or absolutely dis-approved categories. The combination of the two effects makes it easier for us to arrive quickly at positive conclusions, both when moral issues are at stake and when they are not. As we have seen, this is precisely what we do not need. Thus disappearance of the idea of absolute right and wrong will be a step in the right direc-tion. Indeed, it may do much to diminish unreasoning prejudice and increase the likelihood of practical and peaceful solutions to the disputes that constantly arise in today's complex world.

Conclusion

We have seen that the thesis examined in this book—the ultimate reducibility of biology to physical science—has roots extending far into the past. Long-mysterious properties of the heart and circulatory system were explained more than three hundred years ago by ordinary principles of hydraulics; the conversion of inorganic to organic compounds was accounted for in the nineteenth century by the atomic theory of matter; the relationship between the catalytic effectiveness of proteins and the electrical properties of their molecules has been understood for decades. Yet despite such successes, there was at first little inclination to conclude that *all* aspects of biology would ultimately be found to be reducible to physics. For living organisms *seemed* so different from nonliving matter, and their structures and properties were so complicated, as to lend credence to the view that the applicability of the principles of physical science would always be only to the fringe, never to the real core, of biology. Such a conviction in the ultimate separability of biological and physical phenomena also had the great advantage of consistency with the anthropocentric notion placing man above and beyond the workings of natural law designed for the regulation of an impersonal world. Because of our need to feel unique, the idea of the fundamental irreconcilability of life processes with the principles of physical science has always been almost automatically accepted as true, at least until proved false by overwhelming evidence.

The clear inference from the trend of scientific discovery examined in this book is that in the mid-twentieth century, such overwhelming evidence finally exists—that the boundary separating the physically explicable fringe phenomena of living organisms from the nonphysical core has been pushed inward until no core is left. We have traced a possible path of evolutionary development leading to organisms with the structure and metabolism of modern creatures including man, and we have done so strictly in terms of the operation of the ordinary laws of physics in the inert material of the prehistoric earth. We have learned how behavior, even of the kind we call intelligent, will probably be found to be completely accounted for when we have adequately developed our understanding of the properties of the complex switching networks that appear to underlie the operation of brains as well as computers. We have studied evidence that the subjective concomitant of thought and sensation—consciousness—is also a lawful phenomenon, with its content determined in regular and predictable fashion by the structure and electrochemical activity of the brain. We have, in short, followed the development of a strong case for concluding that the origin and properties of the human organism—physical, behavioral, mental, subjective and objective—are completely and in detail the consequence of the normal interaction of the ordinary laws and particles of physics. In such terms, man is a machine.

To those who are for the first time encountering this thesis as a scientific theory rather than as a philosophic postulate, it should be said again that the thesis has not yet been proved beyond all doubt. Mistakes are occasionally made in science, and there has not yet been time for all the critical research discoveries to be backed up in depth by the similar findings of many investigators. The rejection of a few selected experimental results and deductions might permit definition of a class of biological phenomena for which it could be predicted that no adequate physical explanation will ever be found. No doubt some—especially those who are repelled by the idea that man is a machine—will consider this the only acceptable interpretation. But others, like the author, will judge the existing evidence to be compelling. For this group is strongly attracted by the idea of a lawful universe. We find great appeal in the notion that all we can observe or feel is caused by the operation of a single set of inviolable physical laws upon a single set of material particles. This seems to us to be a logical extension of the unbroken chain of brilliant successes of physical science in accounting for one aspect after another of human experience. Therefore, to us, the evidence examined in this book seems right; we believe it easily, despite the fact that the resulting displacement of essentially nonphysical Man by machinelike man requires us to relinquish even the small vestige of claim to human uniqueness left to us by the discoveries of Galileo and Darwin.

Fortunately, we have also been able to conclude that the concept of mechanical man poses no serious threat to society. Of course, in a certain sense this is irrelevant to our discussion, for ultimately the acceptance, like the validity, of a new scientific theory is independent of its social attractiveness. For man is essentially logical, and is not indefinitely able to reject convincing evidence, despite his initial prejudices. But man is also innately social, as we have had occasion to observe. Thus none can avoid an interest in the probable effects on civilization of such developments as those discussed here. The revolution in religion, the preservation of morality, the perseverance of individual ambition, the increased liberality of society toward the individual, the decrease in unreasoning prejudice resulting from the elimination of the mystical idea of absolute right and wrong—these consequences seem important to us. It is therefore comforting to be able to end by restating the conclusion that the net social effect of the resulting changes in attitudes and institutions should in fact be positive rather than negative or neutral. Society profits when its members behave more intelligently. And men who know they are machines should be able to bring a higher degree of objectivity to bear on their problems than machines that think they are Men.

Index